Lecture Notes in Statistics 199

For other titles published in this series, go to
http://www.springer.com/series/694

Hannu Oja

Multivariate Nonparametric Methods with R

An Approach Based on Spatial Signs and Ranks

Springer

Prof. Hannu Oja
University of Tampere
Tampere School of Public Health
FIN-33014 Tampere
Finland
hannu.oja@uta.fi

ISSN 0930-0325
ISBN 978-1-4419-0467-6 e-ISBN 978-1-4419-0468-3
DOI 10.1007/978-1-4419-0468-3
Springer New York Dordrecht Heidelberg London

Library of Congress Control Number: 2010924740

Printed on acid-free paper

Springer is part of Springer Science+Business Media (www.springer.com)

To my family.

Preface

This book introduces a new way to analyze multivariate data. The analysis of data based on multivariate spatial signs and ranks proceeds very much as does a traditional multivariate analysis relying on the assumption of multivariate normality: the L_2 norm is just replaced by different L_1 norms, observation vectors are replaced by their (standardized and centered) spatial signs and ranks, and so on. The methods are fairly efficient and robust, and no moment assumptions are needed. A unified theory starting with the simple one-sample location problem and proceeding through the several-sample location problems to the general multivariate linear regression model and finally to the analysis of cluster-dependent data is presented.

The material is divided into 14 chapters. Chapter 1 serves as a short introduction to the general ideas and strategies followed in the book. Chapter 2 introduces and discusses different types of parametric, nonparametric, and semiparametric statistical models used to analyze the multivariate data. Chapter 3 provides general descriptive tools to describe the properties of multivariate distributions and multivariate datasets. Multivariate location and scatter functionals and statistics and their use is described in detail. Chapter 4 introduces the concepts of multivariate spatial sign, signed-rank, and rank, and shows their connection to certain L_1 objective functions. Also sign and rank covariance matrices are discussed carefully. The first four chapters thus provide the necessary tools to understand the remaining part of the book.

The one-sample location case is treated thoroughly in Chapters 5-8. The book then starts with the familiar Hotelling's T^2 test and the corresponding estimate, the sample mean vector. The spatial sign test with the spatial median as well as the spatial signed-rank test with Hodges-Lehmann estimate is treated in Chapters 6 and 7. All the tests and estimates are made practical; the algorithms for the estimates and the estimates of the covariance matrices of the estimates are also discussed and described in detail. Chapter 7 is devoted to the comparisons of these three competing approaches.

Chapters 9 and 10 continue discussion of the one-sample case. In Chapter 9, tests and estimates for the shape (scatter) matrices based on different sign and rank covariance matrices are given. Also principal component analysis is discussed. Chapter 10 provides different tests for the important hypothesis of independence between two subvectors.

Sign and rank tests with companion estimates for the comparison of two or several treatment effects are given in Chapter 11 (independent samples) and Chapter 12 (randomized block design). The general multivariate multiple regression case with L_1 objective functions is finally discussed in Chapter 13. The book ends with sign and rank procedures for cluster-dependent data in Chapter 14.

Throughout the book, the theory is illustrated with examples. For computation of the statistical procedures described in the book, the R package MNM (and SpatialNP) is available on CRAN. In the analysis we always compare three different score functions, the identity score, the spatial sign score, and the spatial rank (or spatial signed-rank) score, and the general estimating and testing strategy is explained in each case. Some basic vector and matrix algebra tools and asymptotic results are given in Appendices A and B.

Acknowledgements
The research reported in this book is to a great degree based on the thesis work of several ex-students of mine, including Ahti Niinimaa, Jyrki Möttönen, Samuli Visuri, Esa Ollila, Sara Taskinen, Jaakko Nevalainen, Seija Sirkiä, and Klaus Nordhausen. I wish to thank them all. This would not have been possible without their work. I have been lucky to have such excellent students. My special thanks go to Klaus Nordhausen for his hard work in writing and putting together (with Jyrki and Seija) the R-code to implement the theory. I am naturally also indebted to many colleagues and coauthors for valuable and stimulating discussions. I express my sincere thanks for discussions and cooperation in this specific research area with Biman Chakraborty, Probal Chaudhuri, Christopher Croux, Marc Hallin, Tom Hettmansperger, Visa Koivunen, Denis Larocque, Jukka Nyblom, Davy Paindaveine, Ron Randles, Bob Serfling, Juha Tienari, and Dave Tyler.

Thanks are also due to the Academy of Finland for several research grants for working in the area of multivariate nonparametric methods. I also thank the editors of this series and John Kimmel of Springer-Verlag for his encouragement and patience.

Tampere,
January 2010 *Hannu Oja*

Contents

Notation

$\mathbf{Y} = (\mathbf{y}_1, \mathbf{y}_2, ..., \mathbf{y}_n)'$ $n \times p$ matrix of response variables
$\mathbf{X} = (\mathbf{x}_1, \mathbf{x}_2, ..., \mathbf{x}_n)'$ $n \times q$ matrix of explaining variables
$N_p(\mu, \Sigma)$ p-variate normal distribution
$t_{\nu,p}(\mu, \Sigma)$ p-variate t-distribution with ν degrees of freedom
$E_p(\mu, \Sigma, \rho)$ p-variate elliptic distribution
$\mathscr{S}_p$ p-variate unit sphere
$\mathscr{B}_p$ p-variate unit ball
$|\mathbf{A}|$ Matrix norm $\sqrt{tr(\mathbf{A}'\mathbf{A})}$
$\mathbf{M}(F)$ Multivariate location vector (functional)
$\mathbf{C}(F)$ Multivariate scatter matrix (functional)
$\mathbf{M}(\mathbf{Y})$ Multivariate location statistic
$\mathbf{S}(\mathbf{Y})$ Multivariate scatter statistic
$\mathbf{T}(\mathbf{y})$ General score function
$\mathbf{L}(\mathbf{y})$ Optimal location score function
$\mathbf{U}(\mathbf{y})$ Spatial sign function
$\mathbf{R}(\mathbf{y}), \mathbf{R}_F(\mathbf{y})$ Spatial rank function
$\mathbf{Q}(\mathbf{y}), \mathbf{Q}_F(\mathbf{y})$ Spatial signed-rank function
$\mathbf{U}_i = \mathbf{U}(\mathbf{y}_i)$ Spatial sign of $\mathbf{y}_i$
$\mathbf{R}_i = \mathbf{R}(\mathbf{y}_i)$ Spatial rank of $\mathbf{y}_i$
$\mathbf{Q}_i = \mathbf{Q}(\mathbf{y}_i)$ Spatial signed-rank of $\mathbf{y}_i$
AVE Average
COV Covariance matrix
UCOV Sign covariance matrix
TCOV Sign covariance matrix based of pairs of observations
QCOV Signed-rank covariance matrix
RCOV Rank covariance matrix

Chapter 1
Introduction

Classical multivariate statistical inference methods (Hotelling's T^2, multivariate analysis of variance, multivariate regression, tests for independence, canonical correlation analysis, principal component analysis, etc.) are based on the use of the regular sample mean vector and covariance matrix. See, for example, the monographs by Anderson (2003) and Mardia et al. (1979). These standard moment-based multivariate techniques are optimal under the assumption of multivariate normality but unfortunately poor in their efficiency for heavy-tailed distributions and are highly sensitive to outlying observations. The book by Puri and Sen (1971) gives a complete presentation of multivariate analysis methods based on marginal signs and ranks. Pesarin (2001) considers permutation tests for multidimensional hypotheses. In this book nonparametric and robust competitors to standard multivariate inference methods based on (multivariate) spatial signs and ranks are introduced and discussed in detail. An R statistical software package MNM (Multivariate Nonparametric Methods) to implement the procedures is freely available to users of spatial sign and rank methods.

The univariate concepts of sign and rank are based on the ordering of the univariate data $y_1, \ldots, y_n$. The ordering is manifested with the univariate sign function $U(y)$ with values -1, 0, and 1 for $y < 0$, $y = 0$, and $y > 0$, respectively. The sign and centered rank of the observation y_i is then $U(y_i)$ and $\mathbf{AVE}_j\{U(y_i - y_j)\}$. In the multivariate case there is no natural coordinate-free ordering of the data points; see Barnett (1976) for a discussion on the problem. An approach utilizing L_1 objective or criterion functions is therefore often used to extend these concepts to the multivariate case. Let $\mathbf{Y} = (\mathbf{y}_1, \ldots, \mathbf{y}_n)'$ be an $n \times p$ data matrix with n observations and p variables. The multivariate spatial sign $\mathbf{U}_i$, multivariate spatial (centered) rank $\mathbf{R}_i$, and multivariate spatial signed-rank $\mathbf{Q}_i$, $i = 1, \ldots, n$, may be implicitly defined using the three L_1 criterion functions with Euclidean norm $|\cdot|$. The sign, rank and signed-rank are then defined implicitly by

H. Oja, *Multivariate Nonparametric Methods with R: An Approach Based on Spatial Signs and Ranks*, Lecture Notes in Statistics 199, DOI 10.1007/978-1-4419-0468-3_1,

$$\begin{aligned}
\mathbf{AVE}\{|\mathbf{y}_i|\} &= \mathbf{AVE}\{\mathbf{U}_i'\mathbf{y}_i\},\\
\frac{1}{2}\mathbf{AVE}\{|\mathbf{y}_i-\mathbf{y}_j|\} &= \mathbf{AVE}\{\mathbf{R}_i'\mathbf{y}_i\}, \text{ and}\\
\frac{1}{4}\mathbf{AVE}\{|\mathbf{y}_i-\mathbf{y}_j|+|\mathbf{y}_i+\mathbf{y}_j|\} &= \mathbf{AVE}\{\mathbf{Q}_i'\mathbf{y}_i\}.
\end{aligned}$$

See Hettmansperger and Aubuchon (1988). Note also that the sign, centered rank, and signed-rank may be seen as *scores* $\mathbf{T}(\mathbf{y})$ corresponding to the three objective functions. These score functions then are

$$\begin{aligned}
\mathbf{U}(\mathbf{y}) &= |\mathbf{y}|^{-1}\mathbf{y},\\
\mathbf{R}(\mathbf{y}) &= \mathbf{AVE}\{\mathbf{U}(\mathbf{y}-\mathbf{y}_i)\}, \text{ and}\\
\mathbf{Q}(\mathbf{y}) &= \frac{1}{2}\mathbf{AVE}\{\mathbf{U}(\mathbf{y}-\mathbf{y}_i)+\mathbf{U}(\mathbf{y}+\mathbf{y}_i)\}.
\end{aligned}$$

The identity score $\mathbf{T}(\mathbf{y}_i)=\mathbf{y}_i$, $i=1,...,n$, is the score corresponding to the regular L_2 criterion $\mathbf{AVE}\{|\mathbf{y}_i|^2\}=\mathrm{AVE}\{\mathbf{y}_i'\mathbf{y}_i\}$.

Multivariate spatial sign and spatial rank methods are thus based on the L_1 objective functions and corresponding score functions. The L_1 methods have a long history in the univariate case but they are often regarded as computationally highly demanding. See, however, Portnoy and Koenker (1997). The first objective function $\mathbf{AVE}\{|\mathbf{y}_i|\}$, if applied to the residuals in the general linear regression model, is the *mean deviation* of the residuals from the origin, and it is the basis for the so called least absolute deviation (LAD) methods. It yields different median-type estimates and sign tests in the one-sample, two-sample, several-sample and finally general linear model settings. The second objective function $\mathbf{AVE}\{|\mathbf{y}_i-\mathbf{y}_j|\}$ is the *mean difference* of the residuals which in fact measures how close together the residuals are. The second and third objective functions generate Hodges-Lehmann type estimates and rank tests for different location problems.

The general strategy in the analysis of the multivariate data followed in this book is first to replace the original observations $\mathbf{y}_i$ by some scores $\mathbf{T}_i=\mathbf{T}(\mathbf{y}_i)$ or, in more complex designs, by centered and/or standardized scores $\hat{\mathbf{T}}_i=\hat{\mathbf{T}}(\mathbf{y}_i)$, $i=1,...,n$. The statistical tests are then based on the new data matrix

$$\mathbf{T}=(\mathbf{T}_1,...,\mathbf{T}_n)' \quad \text{or} \quad \hat{\mathbf{T}}=(\hat{\mathbf{T}}_1,...,\hat{\mathbf{T}}_n)'.$$

The spatial sign score $\mathbf{U}(\mathbf{y})$, the spatial rank score $\mathbf{R}(\mathbf{y})$, and the spatial signed-rank score $\mathbf{Q}(\mathbf{y})$ thus correspond to the three L_1 criterion functions given above. The tests are then rotation invariant but not affine invariant. Inner centering and/or standardization is used to attain the desired affine invariance property of the tests. The location estimates are chosen to minimize the selected criterion function; they are also obtained if one applies inner centering with the corresponding score. Inner standardization may be used to construct affine equivariant versions of the estimates and as a side product one gets scatter matrix estimates for the inference on the covariance structure of the data.

The tests and estimates for the multivariate location problem based on multivariate signs and ranks have been widely discussed in the literature. See, for example, Möttönen and Oja (1995), Choi and Marden (1997), Marden (1999a), and Oja and Randles (2004). The scatter matrix estimates by Tyler (1987) and Dümbgen (1998) are often used in inner standardizations. The location tests and estimates are robust and they have good efficiency properties even in the multivariate normal model. Möttönen et al. (1997) calculated the asymptotic efficiencies $e_1(p,\nu)$ and $e_2(p,\nu)$ of the multivariate spatial sign and rank methods, respectively, in the p-variate $t_{\nu,p}$ distribution case ($t_{\infty,p}$ is the p-variate normal distribution). In the 3-variate case, for example, the asymptotic efficiencies are

$$e_1(3,3) = 2.162, \quad e_1(3,10) = 1.009, \quad e_1(3,\infty) = 0.849,$$
$$e_2(3,3) = 1.994, \quad e_2(3,10) = 1.081, \quad e_2(3,\infty) = 0.973,$$

and in the 10-variate case one has even higher

$$e_1(10,3) = 2.422, \quad e_1(10,10) = 1.131, \quad e_1(10,\infty) = 0.951,$$
$$e_2(10,3) = 2.093, \quad e_2(10,10) = 1.103, \quad e_2(10,\infty) = 0.989.$$

The procedures based on spatial signs and ranks, however, yield only one possible approach to multivariate nonparametric tests (sign test, rank test) and corresponding estimates (median, Hodges-Lehmann estimate). Randles (1989), for example, developed an affine invariant sign test based on *interdirections*. Interdirections measure the angular distance between two observation vectors relative to the rest of the data. Randles (1989) was followed by a series of papers introducing nonparametric sign and rank interdirection tests for different location problems. These tests are typically asymptotically equivalent with affine invariant versions of the spatial sign and rank tests. The tests and estimates based on interdirections are, unfortunately, computationally heavy.

The multivariate inference methods based on marginal signs and ranks are described in detail in the monograph by Puri and Sen (1971). This first extension of the univariate sign and rank methods to the multivariate setting is based on the criterion functions

$$\mathbf{AVE}\{|y_{i1}| + \cdots + |y_{ip}|\} \text{ and } \mathbf{AVE}\{|y_{i1} - y_{j1}| + \cdots + |y_{ip} - y_{jp}|\},$$

respectively. Note that here the Euclidean norm, the L_2-norm $|\mathbf{y}| = \sqrt{y_1^2 + \cdots + y_p^2}$, is replaced by the L_1-norm $|\mathbf{y}| = |y_1| + \cdots + |y_p|$. Unfortunately, the tests are not affine invariant and the estimates are not affine equivariant in this approach. Chakraborty and Chaudhuri (1996) and Chakraborty and Chaudhuri (1998) used the so-called transformation retransformation technique (TR-technique) to find affine invariant versions of the tests and affine equivariant versions of the estimates.

Affine equivariant multivariate signs and ranks are obtained if one uses the L_1 criterion functions

$$\mathbf{AVE}\{V(\mathbf{y}_{i_1},\ldots,\mathbf{y}_{i_p},\mathbf{0})\} \quad \text{and} \quad \mathbf{AVE}\{V(\mathbf{y}_{i_1},\ldots,\mathbf{y}_{i_{p+1}})\}$$

where the average is over all the p-tuples and $(p+1)$-tuples of observations, respectively, and

$$V(\mathbf{y}_1,\ldots,\mathbf{y}_{p+1}) = \frac{1}{p!}\text{abs}\left\{\det\begin{pmatrix} 1 & \cdots & 1 \\ \mathbf{y}_1 & \cdots & \mathbf{y}_{p+1}\end{pmatrix}\right\}$$

is the volume of the p-variate simplex with vertices $\mathbf{y}_1,\ldots,\mathbf{y}_{p+1}$. The one-sample location estimate is known as the Oja median Oja (1983). The approach in the one-sample and two-sample location cases is described in Oja (1999). In a parallel approach Koshevoy and Mosler (1997a,b, 1998) used so-called zonotopes and lift-zonotopes to illustrate and characterize a multivariate data cloud. The duality relationship between these two approaches is analyzed in Koshevoy et al. (2004).

In all the approaches listed above the power of the rank tests may often be increased, if some further transformations are applied to the ranks. In a series of papers, Hallin and Paindaveine constructed *optimal signed-rank tests* for the location and scatter problems in the *elliptical model*; see the series of papers starting with Hallin and Paindaveine (2002). In their approach, the location tests were based on the spatial signs and optimally transformed ranks of the Euclidean lengths of the standardized observations. As the model of elliptically symmetric distributions, the so called *independent component (IC) model* is also an extension of the multivariate normal model. In the test construction in this model, one first transforms the observations to the estimated independent coordinates, then calculates the values of the marginal sign and rank test statistics, and finally combines the asymptotically independent tests in the regular way. See Nordhausen et al. (2009) and Oja et al. (2009) for optimal rank tests in the IC model.

In this book the approach is thus based on the spatial signs and ranks. A unified theory starting with the simple one-sample location problem and proceeding through the several-sample location problems to the general multivariate linear regression model and finally to the analysis of cluster-dependent data is presented. The theory is often presented using a general score function and, for comparison to classical methods, the classical normal-based approach (L_2 criterion and identity score function) is carefully reviewed in each case. Also statistical inference on scatter or shape matrices based on the spatial signs and ranks is discussed. The theory is illustrated with several examples. For the computation of the statistical procedures described in the book, R packages MNM and SpatialNP are available on CRAN. The readers who are not so familiar with R are advised to learn more from Venables et al. (2009), Dalgaard (2008), or Everitt (2005).

Chapter 2
Multivariate location and scatter models

Abstract In this chapter we first introduce and describe different symmetrical and asymmetrical parametric and semiparametric (linear) models which are then later used as the model assumptions in the statistical analysis. The models discussed include multivariate normal distribution $N_p(\mu,\Sigma)$ and its different extensions including multivariate t distribution $t_{\nu,p}(\mu,\Sigma)$ distribution, multivariate elliptical distribution $E_p(\mu,\Sigma,\rho)$, as well as still wider semiparametric symmetrical models. Also some models with skew distributions (generalized elliptical model, mixture models, skew-elliptical model, independent component model) are briefly discussed.

2.1 Construction of the multivariate models

In this chapter we consider different parametric, nonparametric, and semiparametric models for multivariate continuous observations. Consider a data matrix consisting of n observed values of a p-variate response variable,

$$\mathbf{Y} = (\mathbf{y}_1,...,\mathbf{y}_n)' \quad \in \mathscr{M}(n,p),$$

where $\mathscr{M}(n,p)$ is the set of $n \times p$ matrices. In the one-sample case, the p-variate observations $\mathbf{y}_i$, $i = 1,...,n$, may be thought to be independent and to be generated by

$$\mathbf{y}_i \ = \ \mu \ + \ \Omega \ \varepsilon_i, \quad i = 1,...,n,$$

where the p-vectors ε_i are called *standardized and centered residuals*, μ is a *location p-vector*, Ω is a full-rank $p \times p$ *transformation matrix* and $\Sigma = \Omega\Omega' > 0$ is called a *scatter matrix*. We are more explicit later regarding what is meant by a standardized and centered random vector. (It is usual to say that ε_i is standardized if $\mathbf{COV}(\varepsilon_i) = \mathbf{I}_p$, and it is centered if $\mathbf{E}(\varepsilon_i) = 0$.) Notation $\Sigma > 0$ means that Σ is positive definite (with rank p).

In the several-sample and multivariate regression case, we write

H. Oja, *Multivariate Nonparametric Methods with R: An Approach Based on Spatial Signs and Ranks*, Lecture Notes in Statistics 199, DOI 10.1007/978-1-4419-0468-3_2,

$$\mathbf{X} = (\mathbf{x}_1, ..., \mathbf{x}_n)'$$

for an $n \times q$ matrix of the values of q explaining variables measured on n individuals and assume that

$$\mathbf{Y} = \mathbf{X}\beta + \varepsilon\Omega',$$

where β is the $q \times p$ matrix of regression coefficients and ε is a $n \times p$ matrix of n independent and standardized residuals $\varepsilon_1, ..., \varepsilon_n$. The one-sample case is obtained with $\mathbf{X} = \mathbf{1}_n$. In this book the aim is to develop statistical inference methods, tests and estimates, for unknown parameters β and also for $\Sigma = \Omega\Omega'$ under weak assumptions on the distribution of the standardized residuals ε_i, $i = 1, ..., n$.

Different parametric and semiparametric (or nonparametric) models are obtained by making different assumptions on the distribution of ε_i. First we list symmetry assumptions often needed in future developments. *Symmetry* of a distribution of the standardized p-variate random variable ε may be seen as an invariance property of the distribution under certain transformations. Relevant transformations include *orthogonal transformations* ($\varepsilon \to \mathbf{O}\varepsilon$ with $\mathbf{O}'\mathbf{O} = \mathbf{O}\mathbf{O}' = \mathbf{I}_p$), *sign-change transformations* ($\varepsilon \to \mathbf{J}\varepsilon$, where $\mathbf{J}$ is a sign-change matrix, a $p \times p$ diagonal matrix with diagonal elements ± 1), and *permutations* $\varepsilon \to \mathbf{P}\varepsilon$, where $\mathbf{P}$ is a $p \times p$ permutation matrix (obtained by successively permuting the rows and/or columns of $\mathbf{I}_p$). See also Appendix A for our matrix notation and definitions.

Definition 2.1. The random p-vector ε is

- *spherically symmetrical* if $\mathbf{O}\varepsilon \sim \varepsilon$ for all orthogonal matrices $\mathbf{O}$
- *marginally symmetrical* if $\mathbf{J}\varepsilon \sim \varepsilon$ for all sign-change matrices $\mathbf{J}$
- *symmetrical (or centrally symmetrical)* if $-\varepsilon \sim \varepsilon$
- *exchangeable* if $\mathbf{P}\varepsilon \sim \varepsilon$ for all permutation matrices $\mathbf{P}$

Note that a spherically symmetrical random variable is also marginally symmetrical, centrally symmetrical and exchangeable as $\mathbf{J}$, $-\mathbf{I}_p$ and $\mathbf{P}$ are all orthogonal. A marginally symmetrical random variable is naturally also centrally symmetrical. A hierarchy of symmetrical models is then obtained with assumptions

(A0) $\varepsilon_i \sim N_p(\mathbf{0}, \mathbf{I}_p)$
(A1) ε_i spherically symmetrical
(A2) ε_i marginally symmetrical and exchangeable
(A3) ε_i marginally symmetrical
(A4) ε_i symmetrical

The first and strongest assumption gives the regular multivariate normal (parametric) model with $\mathbf{y}_i \sim N_p(\mu, \Sigma)$ (or $\mathbf{y}_i \sim N_p(\mathbf{x}_i'\beta, \Sigma)$). The classical multivariate inference methods (Hotelling's T^2, multivariate analysis of variance (MANOVA), multivariate regression analysis, principal component analysis (PCA), canonical correlation analysis (CCA), factor analysis (FA), etc.) rely on the assumption of multivariate normality. The multivariate normal distribution is discussed in Section 2.2. In models (A1)–(A4), which may be seen as semiparametric extensions of the multivariate normal model (A0), the location parameter μ is the well defined symmetry center of the distribution of $\mathbf{y}_i$. Extra assumptions are needed to make Σ well defined. In the models (A0)–(A2) the scatter matrix Σ has a natural interpretation; Σ is proportional to the covariance matrix if it exists.

Asymmetrical semiparametric models are obtained if the assumptions are made merely on the distribution of the *direction vectors* $\mathbf{u}_i = |\varepsilon_i|^{-1}\varepsilon_i$. Our directional symmetry assumptions then are

(B1) $\mathbf{u}_i$ uniformly distributed on unit sphere $\mathscr{S}_p$
(B2) $\mathbf{u}_i$ marginally symmetrical and exchangeable
(B3) $\mathbf{u}_i$ marginally symmetrical
(B4) $\mathbf{u}_i$ symmetrical

Note that the assumptions (B1)–(B4) do not say anything about the distribution of the *radius* or *modulus* $r_i = |\varepsilon_i|$ of the standardized vector; the radius r_i and direction vector $\mathbf{u}_i$ may even be dependent so that skew distributions are allowed.

Under the assumptions (B1)–(B4), parameters μ and Σ still describe the location and the scatter of the distribution of the ε_i. Under assumption (B2), for example, the directions $\mathbf{u}_i = |\varepsilon_i|^{-1}\varepsilon_i$ of the transformed (standardized and centered) observations $\varepsilon_i = \Omega^{-1}(\mathbf{y}_i - \mu)$ are "uniformly" distributed in the sense that $\mathbf{E}(\mathbf{u}_i) = \mathbf{0}$ and $p \cdot \mathbf{E}(\mathbf{u}_i\mathbf{u}_i) = \mathbf{I}_p$. (Then μ and Σ are the so called *Hettmansperger-Randles functionals* which are discussed later in the book.) Even in the weakest model (B4), μ is a natural location parameter in the sense that $\mathbf{E}(\mathbf{u}_i) = \mathbf{0}$. (We later show that the *spatial median* of ε_i is then the zero vector.) Also, all hyperplanes going through μ divide the probability mass of the distribution of $\mathbf{y}_i$ into two parts of equal size $1/2$. (Then μ is the so called *half-space median* or *Tukey median*.) Note, however, that in this models the scatter matrix Σ is no longer related to the regular covariance matrix.

In the following, we refer to these models or families of distributions for $\mathbf{y}_i$ using symbols (A0)–(A4) and (B1)–(B4). Then the model (A1) is the so called *elliptical model* which we discuss in more detail in Section 2.2. The model (A2) with somewhat weaker assumptions is called in the following the *location-scatter model*. This model includes all elliptical distributions and, for example, distributions with independent and identically distributed components. It is straightforward to see the following.

Theorem 2.1. *The assumptions (A0)–(A4) and (B1)–(B4) satisfy the joint hierarchy*

$$
\begin{array}{ccccccccc}
(A0) & \Rightarrow & (A1) & \Rightarrow & (A2) & \Rightarrow & (A3) & \Rightarrow & (A4) \\
 & & \Downarrow & & \Downarrow & & \Downarrow & & \Downarrow \\
 & & (B1) & \Rightarrow & (B2) & \Rightarrow & (B3) & \Rightarrow & (B4).
\end{array}
$$

2.2 Multivariate elliptical distributions

In this section we consider the model (A1). As before, let

$$\mathbf{y}_i = \mu + \Omega\, \varepsilon_i, \quad i = 1, ..., n.$$

We assume that $\varepsilon_1, ..., \varepsilon_n$ are independent and identically distributed random vectors from a spherically symmetrical and continuous distribution. We say that the distribution of ε is *spherically symmetrical* around the origin if the density function $f(\varepsilon)$ of ε depends on ε only through the modulus $|\varepsilon|$. We can then write

$$f(\varepsilon) = \exp\{-\rho(|\varepsilon|)\}$$

for some function $\rho(r)$. Note that the equal density contours are then spheres. The modulus $r_i = |\varepsilon_i|$ and direction $\mathbf{u}_i = |\varepsilon_i|^{-1}\varepsilon_i$ are independent, and the direction vector $\mathbf{u}_i$ is uniformly distributed on the p-dimensional unit sphere $\mathscr{S}_p$. It is then easy to see that

$$\mathbf{E}(\mathbf{u}_i) = \mathbf{0} \quad \text{and} \quad \mathbf{COV}(\mathbf{u}_i) = E(\mathbf{u}_i\mathbf{u}_i') = \frac{1}{p}\mathbf{I}_p$$

The density of the modulus is

$$g(r) = c_p r^{p-1} \exp\{-\rho(r)\}, \; r > 0,$$

where

$$c_p = \frac{2\pi^{p/2}}{\Gamma\left(\frac{p}{2}\right)}$$

is the surface area of the unit sphere $\mathscr{S}_p$. The scatter matrix $\Sigma = \Omega\Omega'$ is, however, confounded with g. To fix the scatter matrix Σ one can then, for example, assume that ρ is chosen so that $E(r_i^2) = p$ or $Med(r_i^2) = \chi^2_{p,.5}$ (the median of the chi-square distribution with p degrees of freedom). Then Σ is the regular covariance matrix in the multivariate normal case.

Under these assumptions, the random sample $\mathbf{Y} = (\mathbf{y}_1, ..., \mathbf{y}_n)'$ comes from a p-variate elliptical distribution with probability density function

$$f_\mathbf{y}(\mathbf{y}) = |\Sigma|^{-1/2} f\left(\Sigma^{-1/2}(\mathbf{y} - \mu)\right),$$

where μ is the symmetry center and $\Sigma > 0$ is the scatter matrix (parameter). The matrix $\Sigma^{-1/2}$ is chosen here to be symmetric. The location parameter μ is the mean vector (if it exists) and the scatter matrix Σ is proportional to the regular covariance matrix (if it exists). We also write

$$\mathbf{y}_i \sim E_p(\mu, \Sigma, \rho).$$

Note that the transformation matrix Ω is not uniquely defined in the elliptical model as, for any orthogonal matrix $\mathbf{O}$, $\Omega\varepsilon_i = (\Omega\mathbf{O})(\mathbf{O}'\varepsilon_i) = \Omega^*\varepsilon_i^*$ and also ε_i^* has a spherically symmetric distribution with density f. In principal component analysis the eigenvectors and eigenvalues of Σ are of interest; the orthogonal matrix of eigenvectors $\mathbf{O} = \mathbf{O}(\Sigma)$ and the diagonal matrix of eigenvalues $\mathbf{D} = \mathbf{D}(\Sigma)$ (in a decreasing order of magnitude) are obtained from the eigenvector and eigenvalue decomposition

$$\Sigma = \mathbf{ODO}'.$$

Write $\mathrm{diag}(\Sigma)$ for a diagonal matrix having the same diagonal elements as Σ. Often, the scatter matrix is normalized as

$$[\mathrm{diag}(\Sigma)]^{-1/2}\Sigma[\mathrm{diag}(\Sigma)]^{-1/2}.$$

This is the *correlation matrix* (if it exists). Another way to normalize the scatter matrix is to divide it by a scale parameter.

Definition 2.2. Let Σ be a scatter matrix. Then a *scale parameter* $\sigma^2 = \sigma^2(\Sigma)$ is the scalar-valued function that satisfies

$$\sigma^2(\mathbf{I}_p) = 1 \quad \text{and} \quad \sigma^2(c \cdot \Sigma) = c \cdot \sigma^2(\Sigma).$$

The normalized matrix

$$\sigma^{-2}\Sigma$$

is then the corresponding *shape matrix*.

Possible choices for the scale parameter are, for example, the arithmetic, geometric, and harmonic means of the eigenvalues of Σ, namely

$$tr(\Sigma)/p, \quad \det(\Sigma), \quad \text{and} \quad [tr(\Sigma^{-1})/p]^{-1}.$$

See Paindaveine (2008) for a discussion of the choices of the scale and shape parameters.

Example 2.1. **Multivariate normal distribution.** Assume that ε_i is spherically symmetric and that $r_i^2 \sim \chi_p^2$. (Now $E(r_i^2) = p$.) It then follows that the distribution of ε_i is the *standard multivariate normal* distribution $N_p(\mathbf{0}, \mathbf{I}_p)$ with the (probability) density function

$$f(\varepsilon) = (2\pi)^{-p/2}\exp\left\{-\frac{\varepsilon'\varepsilon}{2}\right\}.$$

We write $\varepsilon_i \sim N_p(\mathbf{0}, \mathbf{I}_p)$. The p components of ε_i are then independent and distributed as $N(0,1)$. In fact, ε_i is spherically symmetric and has independent components if and only if ε_i has a multivariate normal distribution. Finally, the distribution of $\mathbf{y}_i$ is a p-variate normal distribution $N_p(\mu, \Sigma)$ with the density function

$$f_{\mathbf{y}}(\mathbf{y}) = (2\pi)^{-p/2}|\Sigma|^{-1/2}\exp\left\{-\frac{1}{2}(\mathbf{y}-\mu)'\Sigma^{-1}(\mathbf{y}-\mu)\right\}.$$

The distribution of the direction vector is easily obtained in the multivariate normal case as the components of ε_i are independent and $N(0,1)$ distributed. Thus $\varepsilon_{ij}^2 \sim \chi_1^2$ and $\varepsilon_{ij}^2/2 \sim \Gamma(1/2)$. But then

$$\begin{pmatrix} u_{i1}^2 \\ \ldots \\ u_{ip}^2 \end{pmatrix} = \frac{1}{\varepsilon_i'\varepsilon_i} \begin{pmatrix} \varepsilon_{i1}^2 \\ \ldots \\ \varepsilon_{ip}^2 \end{pmatrix}$$

has a so called *Dirichlet distribution* $D_p(1/2,...,1/2)$. See Section 3.3 in Bilodeau and Brenner (1999). As the distribution of $\mathbf{u}_i$ is the same for all spherical distributions, we have the following.

Theorem 2.2. *Let the distribution of a p-variate random vector ε be spherically symmetric around the origin, and let $\mathbf{u} = |\varepsilon|^{-1}\varepsilon$ be the direction vector. Then $(u_1^2,...,u_p^2)$ has the Dirichlet distribution $D_p(1/2,...,1/2)$. Moreover, $\sum_{i=1}^k u_i^2 \sim Beta(k/2,(p-k)/2)$.*

Example 2.2. **Multivariate t distribution.** Assume again that ε_i is spherically symmetric and that $r_i^2/p \sim F(p,\nu)$ (F distribution with p and ν degrees of freedom). Then the distribution of ε_i is the p-variate t distribution with ν degrees of freedom; write $\varepsilon_i \sim t_{\nu,p}$. The density function of ε_i is then

$$f(\varepsilon) = \frac{\Gamma((p+\nu)/2)}{\Gamma(\nu/2)(\pi\nu)^{p/2}} \left(1 + \frac{\varepsilon'\varepsilon}{\nu}\right)^{-(p+\nu)/2}.$$

The components of ε_i are uncorrelated, but not independent, and their marginal distribution is a univariate t_ν distribution. The smaller ν is, the heavier are the tails of the distribution. The expected value exists if $\nu \geq 2$, and the covariance matrix exists for degrees of freedom $\nu \geq 3$. The very heavy-tailed distribution with $\nu = 1$ is called the *multivariate Cauchy distribution*. The multivariate normal distribution is obtained as a limit case as $\nu \to \infty$. The distribution of $\mathbf{y}_i = \mu + \Omega\varepsilon_i$ is denoted by $t_{\nu,p}(\mu,\Sigma)$. If $\varepsilon^* \sim N_p(\mathbf{0},\mathbf{I}_p)$ and $s^2 \sim \chi_\nu^2$ and ε^* and s^2 are independent then

$$\varepsilon = (s^2/\nu)^{-1/2}\varepsilon^* \sim t_{\nu,p}.$$

Example 2.3. **Multivariate power exponential family.** In the so-called multivariate power exponential family, the density of the distribution of ε_i is

$$f(\varepsilon) = k_{p,\nu} \exp\left\{-\frac{|\varepsilon|^{2\nu}}{2}\right\},$$

where

$$k_{p,\nu} = \frac{p\Gamma(p/2)}{\pi^{p/2}\Gamma((2\nu+p)/(2\nu))2^{(2\nu+p)/(2\nu)}}$$

is determined so that the density integrates up to 1. Now $|\varepsilon_i|^{2\nu} \sim \Gamma(1/2, p/(2\nu))$ which can be used to simulate the observations from this flexible parametric model. If $\nu = 1$ then $\varepsilon_i \sim N_p(\mathbf{0}, \mathbf{I}_p)$. The model includes both heavy-tailed ($\nu > 1$) and light-tailed ($\nu < 1$) elliptical distributions. The (heavy-tailed) multivariate double exponential (Laplace) distribution is given by $\nu = 1/2$ and a (light-tailed) multivariate uniform elliptical distribution is obtained as a limit when $\nu \to \infty$. The distribution of $\mathbf{y}_i = \Omega\varepsilon_i + \mu$ is denoted by $PE(\mu, \Sigma, \nu)$. See Gómez et al. (1998).

2.3 Other distribution families

Example 2.4. **Location-scatter model.** (A2) is a wider model than the elliptical model assuming only that the components of ε_i are exchangeable and marginally symmetric; that is,

$$\mathbf{JP}\varepsilon_i \sim \varepsilon_i, \quad \text{for all sign-changes } \mathbf{J} \text{ and permutations } \mathbf{P}.$$

In addition to elliptical distributions, the family includes distributions where the marginal variables are symmetrical, independent, and identically distributed. The model also includes distributions where ε_i has a density of the general form

$$f(\varepsilon) = \exp\{-\rho(||\varepsilon||)\},$$

where $||\cdot||$ is any permutation and sign change invariant metric ($||\mathbf{JP}\varepsilon|| = ||\varepsilon||$ for all $\mathbf{J}$ and $\mathbf{P}$). This is true for the L_α-norm

$$||\varepsilon||_\alpha = (|\varepsilon_1|^\alpha + \cdots + |\varepsilon_p|^\alpha)^{1/\alpha},$$

for example, and therefore a wide variety of distributional shapes is available in this model. Recall that the elliptical model is given with the L_2-norm.

We thus assume that our observed vectors $\mathbf{y}_i$ are generated by $\mathbf{y}_i = \mu + \Omega\varepsilon_i$, $i = 1, ..., n$. As in the elliptic model, both μ and $\Sigma = \Omega\Omega'$ are well defined if we again assume that $E(|\varepsilon_i|^2) = p$ or $Med(|\varepsilon_i|^2) = \chi^2_{p,.5}$. Thus μ is the symmetry center and Σ is proportional to the covariance matrix if it exists. (Σ is the covariance matrix in the multivariate normal case.)

Note that if we still weaken the assumptions and assume only that

$$\mathbf{J}\varepsilon_i \sim \varepsilon_i, \quad \text{for all sign-changes } \mathbf{J},$$

μ is still well defined (symmetry center) but $\Sigma = \Omega\Lambda\Omega'$, where Λ is a diagonal matrix depending on the (unknown) distribution of ε_i. It is, however, still possible to fix some of the features of the transformation matrix Ω.

Example 2.5. **Generalized elliptical model (B1) and its extension (B2).** The definition of the generalized elliptical model was given by Frahm (2004). In this model he assumes that

$$\mathbf{O}\mathbf{u}_i \sim \mathbf{u}_i, \quad \text{for all orthogonal } \mathbf{O}.$$

The direction vectors $\mathbf{u}_i = |\varepsilon_i|^{-1}\varepsilon_i$ are then distributed as if the original observations ε_i were coming from an elliptically symmetric distribution. Randles (2000) considered the same assumption saying that the distribution has "elliptical directions" However, no assumptions on the distribution of modulus $r_i = |\varepsilon_i|$ are made; skew distributions may be obtained if the distribution of r_i depends on $\mathbf{u}_i$. Again, a weaker model is obtained if

$$\mathbf{J}\mathbf{P}\mathbf{u}_i \sim \mathbf{u}_i, \quad \text{for all sign-changes } \mathbf{J} \text{ and permutations } \mathbf{P}.$$

Still in this wider model, the *spatial median* of ε_i is $\mathbf{0}$ and the so-called *Tyler's scatter matrix* (discussed later) is proportional to the identity matrix. This fixes location and shape of the distribution of $\mathbf{y}_i$, that is, parameters μ and Σ (up to scale).

Example 2.6. **Independent component model.** In the independent component model the independent and identically distributed random variables $\mathbf{y}_i$ are generated by $\mathbf{y}_i = \mu + \Omega\varepsilon_i$ where the standardized and centered p-vector ε_i has independent components. In the so-called independent component analysis (ICA) the aim is, based on the data $\mathbf{Y} = (\mathbf{y}_1, ..., \mathbf{y}_n)'$, to find an estimate of a retransformation matrix Ω^{-1} (or transformation matrix Ω) and then transform the data to independent components for further analysis. See Hyvärinen et al. (2001) for the ICA problem. It is then often assumed that only a few of the independent components are informative (dimension reduction problem). The problem is not well posed in this general formulation of the model: if Ω^{-1} transforms to independent coordinates then so do $\mathbf{D}\mathbf{P}\Omega^{-1}$ for all diagonal matrices $\mathbf{D}$ with nonzero diagonal elements and for all permutation matrices $\mathbf{P}$. See Nordhausen et al. (2009b) for additional assumptions needed to make Ω unique.

Example 2.7. **Mixtures of multivariate normal distributions and mixtures of elliptical distribution.** In the mixture model of two multivariate normal distributions, ε_i is thought to come from an $N_p(\mu_1, \Sigma_1)$-distribution with probability π_1 and from an $N_p(\mu_2, \Sigma_2)$-distribution with probability $\pi_2 = 1 - \pi_1$. The parameters (π_1, π_2), (μ_1, μ_2), and (Σ_1, Σ_2) then together determine the location, scatter, skewness, and kurtosis properties of the multivariate distribution. A natural way to standardize and center ε_i is to require that $\mathbf{E}(\varepsilon_i) = \mathbf{0}$ and $\mathbf{COV}(\varepsilon_i) = \mathbf{I}_p$; that is,

$$\pi_1\mu_1 + \pi_2\mu_2 = \mathbf{0} \quad \text{and} \quad \pi_1(\Sigma_1 + \mu_1\mu_1') + \pi_2(\Sigma_2 + \mu_2\mu_2') = \mathbf{I}_p.$$

If $\mu_1 = \mu_2 = \mathbf{0}$ then ε_i is elliptically distributed and the heaviness of the tails is determined by (π_1, π_2) and (Σ_1, Σ_2) Intuitively, if π_2 is "small" and Σ_2 is "large" then the second population may be thought to represent the population where the outliers originated. If, for example, $\mathbf{\Sigma}_1 = \mathbf{I}_p$ and $\mathbf{\Sigma}_2 = \sigma^2\mathbf{I}_p$ then the probability density function of ε_i is

$$f(\varepsilon) = \pi_1 \cdot (2\pi)^{-p/2} \exp\left\{-\frac{\varepsilon'\varepsilon}{2}\right\} + \pi_2 \cdot (2\pi\sigma^2)^{-p/2} \exp\left\{-\frac{\varepsilon'\varepsilon}{2\sigma^2}\right\}.$$

This mixture of two distributions can be easily extended to the case of general $k \geq 2$ mixtures, and also to the case where the observations come from other elliptical distributions. See McLachlan and Peel (2000).

Example 2.8. **Skew-normal and skew-elliptical model.** In the skew-normal model, the p-variate "standardized" observations ε_i are obtained as follows. Let ε_i^* come from a $(p+1)$-variate standard normal distribution $N_{p+1}(\mathbf{0}, \mathbf{I}_{p+1})$ and write

$$\varepsilon_i = U(\varepsilon_{i,p+1}^* - \alpha - \beta\varepsilon_{ip}^*) \begin{pmatrix} \varepsilon_{i1}^* \\ \cdots \\ \varepsilon_{ip}^* \end{pmatrix}, \quad i = 1, ..., n.$$

Here $U(y) = -1, 0, 1$ if $y <, =, > 0$. Then again $\mathbf{y}_i = \mu + \Omega\varepsilon_i$, $i = 1, ..., n$. The multivariate skew-elliptical distribution may be defined in the same way just by assuming that ε_i^* come from $E_{p+1}(\mathbf{0}, \mathbf{I}_{p+1}, \rho)$. Note that a still more general model is obtained if we assume that

$$\varepsilon_i = s_i \begin{pmatrix} \varepsilon_{i1}^* \\ \cdots \\ \varepsilon_{ip}^* \end{pmatrix},$$

where ε_i^* is coming from $E_p(\mathbf{0}, \mathbf{I}_p, \rho)$ and s_i is a random variable with possible values ± 1 possibly depending on ε_i^*. Note that if $\mu = \mathbf{0}$ is known then $\mathbf{\Sigma}$ is proportional to the regular covariance matrix with respect to the origin; that is $\mathbf{E}(\mathbf{y}_i\mathbf{y}_i') \propto \Sigma$. See Azzalini (2005) and references therein.

Chapter 3
Location and scatter functionals and sample statistics

Abstract In this chapter, we introduce multivariate location and scatter functionals, $\mathbf{M}(F)$ and $\mathbf{S}(F)$, which can be used to describe the multivariate distribution. The sample versions of the functionals, $\mathbf{M}(\mathbf{Y})$ and $\mathbf{S}(\mathbf{Y})$, respectively, can then often be used to estimate the population parameters μ and Σ in the semiparametric models introduced in Chapter 2. Some general properties and the use of the sample statistics are discussed.

3.1 Location and scatter functionals

The characteristics of univariate distributions most commonly considered are location, scale, skewness, and kurtosis. These concepts are often identified with the mean, standard deviation, and standardized third and fourth moments. Skewness and kurtosis are often seen only as secondary statistics indicating the stability of the primary statistics, location and scale. Skewness and kurtosis are also used in (parametric) model selection. In parametric or semiparametric models one often has natural parameters for location and scale, that is, μ and Σ. In wide nonparametric models functionals for location and scale must be used instead. The functionals or measures are then supposed to satisfy certain natural equivariance or invariance properties.

We first define what we mean by a location vector and a scatter matrix defined as vector- and matrix-valued functionals in wide nonparametric families of multivariate distributions (often including discrete distributions as well). Let $\mathbf{y}$ be a p-variate random variable with cumulative distribution function (cdf) $F_{\mathbf{y}}$.

Definition 3.1.
(i) A p-vector $\mathbf{M}(F)$ is a *location vector* (functional) if it is affine equivariant; that is,

$$\mathbf{M}(F_{\mathbf{Ay}+\mathbf{b}}) = \mathbf{AM}(F_{\mathbf{y}}) + \mathbf{b}$$

H. Oja, *Multivariate Nonparametric Methods with R: An Approach Based on Spatial Signs and Ranks*, Lecture Notes in Statistics 199, DOI 10.1007/978-1-4419-0468-3_3,

for all random vectors $\mathbf{y}$, all full-rank $p \times p$-matrices $\mathbf{A}$ and all p-vectors $\mathbf{b}$.
(ii) A symmetric $p \times p$ matrix $\mathbf{S}(F) \geq 0$ is a *scatter matrix* (functional) if it is affine equivariant in the sense that

$$\mathbf{S}(F_{\mathbf{Ay+b}}) = \mathbf{A}\mathbf{S}(F_{\mathbf{y}})\mathbf{A}'$$

for all random vectors $\mathbf{y}$, all full-rank $p \times p$-matrices $\mathbf{A}$ and all p-vectors $\mathbf{b}$.

Classical location and scatter functionals, namely the *mean vector* $\mathbf{E}(\mathbf{y})$ and the *covariance matrix*

$$\mathbf{COV}(\mathbf{y}) = \mathbf{E}\left[(\mathbf{y} - \mathbf{E}(\mathbf{y}))(\mathbf{y} - \mathbf{E}(\mathbf{y}))'\right],$$

serve as first examples. Note that the matrix of second moments $\mathbf{E}(\mathbf{yy}')$, for example, is not a scatter matrix in the regular sense but can be seen as a scatter matrix with respect to the origin.

The theory of location and scatter functionals has been developed mainly to find new tools for robust estimation of the regular mean vector and covariance matrix in a neighborhood of the multivariate normal model or in the wider model of elliptically symmetric distributions. The competitors of the regular covariance matrix do not usually have the so called *independence property*: if random vector $\mathbf{y}$ has independent components then $\mathbf{S}(F_{\mathbf{y}})$ is a diagonal matrix. It is easy to see that the regular covariance matrix has the independence property. Naturally this property is not important in the elliptic model as the multivariate normal distribution is the only elliptical distribution that can have independent margins. On the other hand, the independence property is crucial if one is working in the independent component model mentioned in Chapter 2 (the ICA problem).

Using the affine equivariance properties, one easily gets the following.

Theorem 3.1. *Assume that random vector* $\mathbf{y}$ *is distributed in the location-scatter model (A2); that is,*

$$\mathbf{y} = \mu + \Omega\varepsilon,$$

where

$$\mathbf{JP}\varepsilon \sim \varepsilon, \quad \textit{for all sign-changes } \mathbf{J} \textit{ and permutations } \mathbf{P},$$

μ *is the location vector (parameter) and* $\Sigma = \Omega\Omega'$ *is the scatter matrix (parameter). Then*

(i) $\mathbf{M}(F_{\mathbf{y}}) = \mu$ *for all location vectors* $\mathbf{M}$.
(ii) $\mathbf{S}(F_{\mathbf{Y}}) \propto \Sigma$ *for all scatter matrices* $\mathbf{S}$.

The determinant $\det(\mathbf{S})$ or trace $tr(\mathbf{S})$ is often used as a global measure of multivariate scatter. In fact, $[\det(\mathbf{S})]^{1/p}$ is the geometric mean and $tr(\mathbf{S})/p$ the arithmetic mean of the eigenvalues of $\mathbf{S}$. The functional $\det(\mathbf{COV}(\mathbf{y}))$ is sometimes called the *generalized variance*; the functional $tr(\mathbf{COV}(\mathbf{y}))$ may be seen as a multivariate extension of the variance as well because $tr(\mathbf{COV}(\mathbf{y})) = \mathbf{E}(|\mathbf{y}-\mathbf{E}(\mathbf{y})|^2)$.

There are several alternative competing techniques to construct location and scatter functionals, for example, M-functionals, S-functionals, τ-functionals, projection-based functionals, CM- and MM-functionals, and so on. These functionals and related estimates are discussed throughout in numerous research and review papers, see, fro example, Maronna (1976), Davies (1987), Lopuhaä (1989), and Tyler (2002). See also the recent monograph by Maronna et al. (2006). A common feature in these approaches is that the functionals and related estimates are built for inference in elliptical models only. Next we consider M-functionals in more detail.

Definition 3.2. Location and scatter M-functionals are functionals $\mathbf{M} = \mathbf{M}(F_{\mathbf{y}})$ and $\mathbf{S} = \mathbf{S}(F_{\mathbf{y}})$ which simultaneously satisfy two implicit equations

$$\mathbf{M} = [\mathbf{E}[w_1(r)]]^{-1}\mathbf{E}[w_1(r)\mathbf{y}]$$

and

$$\mathbf{S} = [\mathbf{E}[w_3(r)]]^{-1}\mathbf{E}[w_2(r)(\mathbf{y}-\mathbf{M})(\mathbf{y}-\mathbf{M})']$$

for some suitably chosen weight functions $w_1(r)$, $w_2(r)$, and $w_3(r)$. The random variable r is the Mahalanobis distance between $\mathbf{y}$ and $\mathbf{M}$; that is,

$$r = |\mathbf{y}-\mathbf{M}|_{\mathbf{S}} = \sqrt{(\mathbf{y}-\mathbf{M})'\mathbf{S}^{-1}(\mathbf{y}-\mathbf{M})}.$$

Consider an elliptic model with known $\rho(r)$ and its derivative function $\psi(r) = \rho'(r)$. If one then chooses $w_1(r) = w_2(r) = \psi(r)/r$ and $w_3(r) \equiv 1$, the M-functionals are called the *pseudo maximum likelihood (ML)* functionals corresponding to that specific distribution determined by ρ. In the multivariate normal case $w_1(r) \equiv w_2(r) \equiv 1$, and the corresponding functionals are the mean vector and the covariance matrix again. A classical M-estimate is *Huber's M-functional* with choices

$$w_1(r) = \min(c/r, 1) \quad \text{and} \quad w_2(r) = d \cdot \min(c^2/r^2, 1)$$

(and $w_3(r) \equiv 1$) with some positive tuning constants c and d. The value of the functional does not depend strongly on the tails of the distribution; the tuning constant c controls this property. The constant d is just a scaling factor.

If $\mathbf{M}_1(F)$ and $\mathbf{S}_1(F)$ are any affine equivariant location and scatter functionals then so are the *one-step M-functionals*, starting from $\mathbf{M}_1$ and $\mathbf{S}_1$, and given by

$$\mathbf{M}_2 = [\mathbf{E}[w_1(r)]]^{-1}\mathbf{E}[w_1(r)\mathbf{y}]$$

and

$$\mathbf{S}_2 = \mathbf{E}\left[w_2(r)(\mathbf{y}-\mathbf{M}_1)(\mathbf{y}-\mathbf{M}_1)'\right],$$

where now $r = |\mathbf{y} - \mathbf{M}_1|_{\mathbf{S}_1}$. It is easy to see that $\mathbf{M}_2$ and $\mathbf{S}_2$ are affine equivariant location and scatter functionals as well. Repeating this step until it converges yields the regular M-estimate (with $w_3(r) \equiv 1$).

3.2 Location and scatter statistics

In this section we introduce sample statistics to estimate unknown location and scatter parameters μ and Σ in different models (or the unknown theoretical or population values of location and scatter functionals $\mathbf{M}(\mathbf{F})$ and $\mathbf{S}(\mathbf{F})$). The values of the sample statistics are similarly denoted by $\mathbf{M}(\mathbf{Y})$ and $\mathbf{S}(\mathbf{Y})$, where $\mathbf{Y} = (\mathbf{y}_1, ..., \mathbf{y}_n)' \in \mathcal{M}(n,p)$ (sample space).

Definition 3.3. (i) A p-vector $\mathbf{M}(\mathbf{Y})$ is a *location statistic* if it is affine equivariant in the sense that

$$\mathbf{M}(\mathbf{Y}\mathbf{A}' + \mathbf{1}\mathbf{b}') = \mathbf{A}\mathbf{M}(\mathbf{Y}) + \mathbf{b}$$

for all data matrices $\mathbf{Y}$, all full-rank $p \times p$-matrices $\mathbf{A}$, and all p-vectors $\mathbf{b}$.
(ii) A symmetric $p \times p$ matrix $\mathbf{S}(\mathbf{Y}) \geq 0$ is a *scatter statistic* if it is affine equivariant in the sense that

$$\mathbf{S}(\mathbf{Y}\mathbf{A}' + \mathbf{1}\mathbf{b}') = \mathbf{A}\mathbf{S}(\mathbf{Y})\mathbf{A}'$$

for all datasets $\mathbf{Y}$ with $\mathbf{S}(\mathbf{Y}) > 0$, all full-rank $p \times p$-matrices $\mathbf{A}$, and all p-vectors $\mathbf{b}$.

If $\mathbf{Y} = (\mathbf{y}_1, ..., \mathbf{y}_n)'$ is a random sample, it is then often natural that location and scatter statistics are invariant in the following sense.

Definition 3.4.
(i) A location statistic $\mathbf{M}$ is permutation invariant if $\mathbf{M}(\mathbf{Y}) = \mathbf{M}(\mathbf{P}\mathbf{Y})$ for all $\mathbf{Y}$ and all $n \times n$ permutation matrices $\mathbf{P}$.
(ii) A scatter statistic $\mathbf{S}$ is permutation invariant if $\mathbf{S}(\mathbf{Y}) = \mathbf{S}(\mathbf{P}\mathbf{Y})$ for all $\mathbf{Y}$ and all $n \times n$ permutation matrices $\mathbf{P}$.

Note that if $\mathbf{M}(\mathbf{Y})$ and $\mathbf{S}(\mathbf{Y})$ are not permutation invariant, invariant estimates (with the same bias but with smaller variation; use the Rao-Blackwell theorem) can be easily obtained as

$$\mathbf{AVE_P}\{\mathbf{M}(\mathbf{P}\mathbf{Y})\} \quad \text{and} \quad \mathbf{AVE_P}\{\mathbf{S}(\mathbf{P}\mathbf{Y})\}$$

where the average is over all $n!$ possible permutation matrices $\mathbf{P}$.

In the one-sample location test constructions we often use scatter statistics $\mathbf{S}$ with respect to the origin that are permutation and sign-change invariant, that is,

$$\mathbf{S}(\mathbf{J}\mathbf{P}\mathbf{Y}\mathbf{A}') = \mathbf{A}\mathbf{S}(\mathbf{Y})\mathbf{A}'$$

for all $\mathbf{Y}$ with $\mathbf{S}(\mathbf{Y}) > 0$, all $p \times p$ matrices $\mathbf{A}$, all $n \times n$ permutation matrices $\mathbf{P}$, and all sign-change matrices $\mathbf{J}$.

Location and scatter functionals $\mathbf{M}(F)$ and $\mathbf{S}(F)$ yield corresponding sample statistics simply by applying the definition to a discrete random variable F_n,

$$\mathbf{M}(\mathbf{Y}) = \mathbf{M}(F_n) \quad \text{and} \quad \mathbf{S}(\mathbf{Y}) = \mathbf{S}(F_n),$$

where F_n is the empirical p-variate cumulative distribution function based on sample $\mathbf{Y} = (\mathbf{y}_1, ..., \mathbf{y}_n)'$. These statistics are then automatically permutation invariant.

The different techniques to construct location and scatter functionals (M-functionals, S-functionals, MM-functionals, τ-functionals, etc.) as mentioned in Section 3.1 may thus be used to find the corresponding sample statistics. The M-statistics, for example, may be defined in the following way.

Definition 3.5. Location and scatter M-statistics $\mathbf{M} = \mathbf{M}(\mathbf{Y})$ and $\mathbf{S} = \mathbf{S}(\mathbf{Y})$ simultaneously satisfy two implicit equations

$$\mathbf{M} = [\mathbf{AVE}[w_1(r_i)]]^{-1}\, \mathbf{AVE}\,[w_1(r_i)\mathbf{y}_i]$$

and

$$\mathbf{S} = [\mathbf{AVE}[w_3(r_i)]]^{-1}\, \mathbf{AVE}\left[w_2(r_i)(\mathbf{y}_i - \mathbf{M})(\mathbf{y}_i - \mathbf{M})'\right]$$

for weight functions $w_1(r)$, $w_2(r)$, and $w_3(r)$. The scalar r_i is the Mahalanobis distance between $\mathbf{y}_i$ and M; that is, $r_i = |\mathbf{y}_i - \mathbf{M}|_{\mathbf{S}}$.

3.3 First and second moments of location and scatter statistics

Why do we need different location and scatter functionals and statistics? In symmetrical models (A0)–(A4) all location statistics $\mathbf{M}(\mathbf{Y})$ estimate the symmetry center μ. In models (A0)–(A2) all scatter statistics $\mathbf{S}(\mathbf{Y})$ estimate the same population quantity (up to a scaling factor). The statistical properties (convergence, limiting distribution, efficiency, robustness, computational convenience, etc.) of the estimates may considerably differ, however. One can then just pick up an estimate that is best for his or her purposes.

To consider the possible bias and the accuracy of location and scatter statistics we next find a general structure for their first and second moments for random samples coming from an elliptical distribution (A1) as well as from a location-scatter model (A2). For most results in this section, we refer to Tyler (1982).

In the following, it is notationally easier to work with the vectors rather than with the matrices. The "vec" operation is used to vectorize a matrix. If $\mathbf{S} > 0$ is a scatter matrix then vec(**S**) is a vector obtained by stacking the columns of **S** on top of each other:

$$\mathrm{vec}(\mathbf{S}) = \mathrm{vec}\left((\mathbf{s}_1,\dots,\mathbf{s}_p)\right) = \begin{pmatrix} \mathbf{s}_1 \\ \vdots \\ \mathbf{s}_p \end{pmatrix}.$$

Moreover for treating the $p^2 \times p^2$ covariance matrices of the vectorized $p \times p$ scatter matrices, the following matrices prove very useful for bookkeeping. Let $\mathbf{e}_i$ be a p-vector with the ith element one and others zero, $i = 1,\dots,p$. Then $\sum_{i=1}^{p} \mathbf{e}_i\mathbf{e}_i' = \mathbf{I}_p$ and we write

$$\mathbf{D}_{p,p} = \sum_{i=1}^{p} (\mathbf{e}_i\mathbf{e}_i') \otimes (\mathbf{e}_i\mathbf{e}_i'),$$

$$\mathbf{J}_{p,p} = \sum_{i=1}^{p}\sum_{j=1}^{p} (\mathbf{e}_i\mathbf{e}_j') \otimes (\mathbf{e}_i\mathbf{e}_j'),$$

$$\mathbf{K}_{p,p} = \sum_{i=1}^{p}\sum_{j=1}^{p} (\mathbf{e}_i\mathbf{e}_j') \otimes (\mathbf{e}_j\mathbf{e}_i'), \quad \text{and}$$

$$\mathbf{I}_{p,p} = \sum_{i=1}^{p}\sum_{j=1}^{p} (\mathbf{e}_i\mathbf{e}_i') \otimes (\mathbf{e}_j\mathbf{e}_j').$$

Naturally $\mathbf{I}_{p,p} = \mathbf{I}_{p^2}$.

The covariance structure of a scatter matrix **C** in symmetric models (A1) and (A2) can be formulated using three orthogonal projection matrices

$$\mathbf{P}_1 = \frac{1}{2}(\mathbf{I}_{p,p} + \mathbf{K}_{p,p}) - \mathbf{D}_{p,p}, \quad \mathbf{P}_2 = \mathbf{D}_{p,p} - \frac{1}{p}\mathbf{J}_{p,p}, \quad \text{and} \quad \mathbf{P}_3 = \frac{1}{p}\mathbf{J}_{p,p}.$$

Then, for any $p \times p$ matrix **A**,

1. $\mathbf{P}_1$ collects the off-diagonal elements of symmetrized **A**:

$$\mathbf{P}_1\mathrm{vec}(\mathbf{A}) = \mathrm{vec}\left(\frac{1}{2}(\mathbf{A} + \mathbf{A}') - \mathrm{diag}(\mathbf{A})\right).$$

2. $\mathbf{P}_2$ picks up the centered diagonal elements; that is,

$$\mathbf{P}_2\mathrm{vec}(\mathbf{A}) = \mathrm{vec}\left(\mathrm{diag}(\mathbf{A}) - \frac{tr(\mathbf{A})}{p}(\mathbf{I}_p)\right).$$

3. $\mathbf{P}_3$ projects matrix **A** to the space spanned by the identity matrix:

$$\mathbf{P}_3\mathrm{vec}(\mathbf{A}) = \frac{tr(\mathbf{A})}{p}\mathrm{vec}(\mathbf{I}_p).$$

Consider the location-scatter model (A2) with the assumption that the standardized variable ε_i is marginally symmetrical and exchangeable. Then the following lemma yields a general structure for the first and second moments of location and scatter statistics, $\mathbf{M} = \mathbf{M}(\varepsilon)$ and $\mathbf{S} = \mathbf{S}(\varepsilon)$ calculated for $\varepsilon = (\varepsilon_1, ..., \varepsilon_n)'$. These statistics then clearly satisfy

$$\mathbf{PJM} \sim \mathbf{M} \quad \text{and} \quad \mathbf{PJSJP}' \sim \mathbf{S}, \quad \text{for all } \mathbf{P} \text{ and } \mathbf{J}.$$

Theorem 3.2. *Assume that the p-variate random vector* $\mathbf{M}$ *satisfies* $\mathbf{PJM} \sim \mathbf{M}$ *for all* $p \times p$ *permutation matrices* $\mathbf{P}$ *and all sign-change matrices* $\mathbf{J}$. *Then there is a positive constant* σ^2 *such that*

$$\mathbf{E(M)} = \mathbf{0} \quad \textit{and} \quad \mathbf{COV(M)} = \sigma^2 \mathbf{I}_p.$$

Second, assume that the random symmetric $p \times p$ *matrix* $\mathbf{S} > 0$ *satisfies* $\mathbf{PJSJP}' \sim \mathbf{S}$ *for all* $p \times p$ *permutation matrices* $\mathbf{P}$ *and all sign-change matrices* $\mathbf{J}$. *Then there are positive constants* η, τ_1, *and* τ_2 *and a constant* τ_3 *such that*

$$\mathbf{E(S)} = \eta \mathbf{I} \quad \textit{and} \quad \mathbf{COV}(vec(\mathbf{S})) = \tau_1 \mathbf{P}_1 + \tau_2 \mathbf{P}_2 + \tau_3 \mathbf{P}_3,$$

where

$$\tau_1 = 2 \cdot Var(S_{12}), \; tau_2 = Var(S_{11}) - Cov(S_{11}, S_{22}) \; \textit{and} \; \tau_3 = (p-1) \cdot Cov(S_{11}, S_{22}).$$

Here $Cov(S_{11}, S_{22})$ *means the covariance between any two different diagonal elements of* $\mathbf{S}$.

Proof. As $\mathbf{E(M)} = \mathbf{PJ\,E(M)}$ and $\mathbf{COV(M)} = \mathbf{PJ\,COV(M)\,JP}'$, one easily sees that $\mathbf{E(T)} = \mathbf{0}$ and $\mathbf{COV(M)} = \sigma^2 \mathbf{I}_p$ for some $\sigma^2 > 0$. Similarly, it is straightforward to see that $\mathbf{E(S)} = \eta \mathbf{I}_p$ for some $\eta > 0$. A general formula for the covariance matrix of the vectorized S is

$$\mathbf{COV}(\text{vec}(\mathbf{S})) = \sum_{i=1}^{p} \sum_{j=1}^{p} \sum_{r=1}^{p} \sum_{r=1}^{p} Cov(S_{ij}, S_{rs}) (\mathbf{e}_i \mathbf{e}_r') \otimes (\mathbf{e}_j \mathbf{e}_s').$$

By symmetry arguments, all the variances of the diagonal elements of $\mathbf{S}$ must be the same, say $Var(S_{11})$, and all the variances of the off-diagonal elements must be the same as well, say $Var(S_{12})$, and all the correlations between off-diagonal elements and other elements of $\mathbf{S}$ must be zero. Finally, all the covariances between diagonal elements must also be the same, say $Cov(S_{11}, S_{22})$. The details in the proof are left to the reader.

Under the stronger assumption that $\mathbf{OSO}' \sim \mathbf{S}$ for all orthogonal $p \times p$ matrices $\mathbf{O}$, corresponding to the elliptic model (A1), one can further show that

$$Cov(S_{11}, S_{22}) = Var(S_{11}) - 2 \cdot Var(S_{12})$$

and one gets a still simpler structure for the covariance matrix of the scatter statistic $\mathbf{S}$.

Corollary 3.1. *Assume that the random symmetric $p \times p$ matrix $\mathbf{S} > 0$ satisfies $\mathbf{OSO}' \sim \mathbf{S}$ for all $p \times p$ orthogonal matrices $\mathbf{O}$. Then there are positive constants η and τ_1 and a constant τ_3 such that*

$$\mathbf{E}(\mathbf{S}) = \eta \mathbf{I} \quad \textit{and} \quad \mathbf{COV}(vec(\mathbf{S})) = \tau_1(\mathbf{P}_1 + \mathbf{P}_2) + \tau_3 \mathbf{P}_3,$$

where τ_1 and τ_3 are as in Theorem 3.2.

As mentioned before, scatter matrices are often standardized with transformation $\mathbf{S} \to [p/tr(\mathbf{S})]\mathbf{S}$; the standardized matrix is then called a *shape matrix* (scale information is lost). The covariance structure of the shape matrix in the elliptical case may be found using the following.

Corollary 3.2. *Assume that the random symmetric $p \times p$ matrix $\mathbf{S} > 0$ satisfies $tr(\mathbf{S}) = p$ and $\mathbf{OSO}' \sim \mathbf{S}$ for all $p \times p$ orthogonal matrices $\mathbf{O}$. Then there is a positive constant τ_1 such that*

$$\mathbf{E}(\mathbf{S}) = \mathbf{I} \quad \textit{and} \quad \mathbf{COV}(vec(\mathbf{S})) = \tau_1(\mathbf{P}_1 + \mathbf{P}_2),$$

where, as before, $\tau_1 = 2 \cdot Var(S_{12})$.

In the general elliptical case we then obtain the following using Theorem 3.2 and Corollary 3.1 and the affine equivariance property of the location and scatter statistics.

Theorem 3.3. *Assume that $\mathbf{Y}$ is a random sample from an elliptical distribution with location vector μ and scatter matrix Σ. For any location and scatter statistics $\mathbf{M}$ and $\mathbf{S}$, there are positive constants η, σ^2, and τ_1 and a constant τ_3 such that*

$$\mathbf{E}(\mathbf{M}(\mathbf{Y})) = \mu \quad \textit{and} \quad \mathbf{COV}(\mathbf{M}(\mathbf{Y})) = \sigma^2 \Sigma$$

and $\mathbf{E}(\mathbf{S}(\mathbf{Y})) = \eta \Sigma$ and

$$\mathbf{COV}(vec(\mathbf{S}(\mathbf{Y}))) = \tau_1(\mathbf{P}_1 + \mathbf{P}_2)\Sigma \otimes \Sigma + \tau_3 vec(\Sigma)(vec(\Sigma))'.$$

Theorem 3.3 thus implies that, in the elliptical model, all location statistics are unbiased estimators of the symmetry center μ. The constant σ^2 may then be used in efficiency comparisons. Remember, however, that σ^2 for the choice $\mathbf{M}$ depends on the distribution of $r_i = |\mathbf{z}_i|$ and on the sample size n. The scatter statistic $\mathbf{S}$ is unbiased for $\eta\Sigma$, η again depending on $\mathbf{S}$, the sample size n, and the distribution of r_i. Then the *correction factors* may be used to guarantee the unbiasedness (or at least consistency) in the case of multivariate normality, for example. The constants τ_1 and τ_3 determine the variance-covariance structure of a scatter estimate $\mathbf{S}$ for sample size n and distribution $F_{\mathbf{y}_i}$. In many multivariate procedures based on the covariance matrix, it is sufficient to know the covariance matrix only up to a constant, that is, the shape matrix. According to Corollary 3.2, the constant τ_1 is sufficient for shape matrix efficiency comparisons. The problem, of course, is how to estimate these constants which are unknown in practice.

3.4 Breakdown point

The breakdown point (BP) of a sample statistic $\mathbf{T} = \mathbf{T}(\mathbf{Y})$ is a measure of *global robustness*. We discuss here the breakdown properties of a sample statistic only, in the spirit of Donoho and Huber (1983). Hampel (1968) was the first one to consider the breakdown point of a functional $\mathbf{T}(F)$ in an asymptotic sense.

Roughly speaking, the breakdown point of a statistic $\mathbf{T}$ is the maximum proportion of contaminated "bad" observations in a dataset $\mathbf{Y}$ that can make the observed value $\mathbf{T}(\mathbf{Y})$ totally unreliable or uninformative. To be more specific, define the m neighborhood $B_m(\mathbf{Y})$ of $\mathbf{Y}$ in sample space $\mathscr{M}(n,p)$ as

$$B_m(\mathbf{Y}) = \{\mathbf{Y}^* \in \mathscr{M}(n,p) \ : \ \mathbf{Y} \text{ and } \mathbf{Y}^* \text{ have } n-m \text{ joint rows}\},$$

$m = 0,...,n$. A corrupted dataset $\mathbf{Y}^* \in B_m(\mathbf{Y})$ may thus be obtained by replacing m observation vectors (rows) in $\mathbf{Y}$ with arbitrary values. Clearly $B_0 \subset B_1 \subset \cdots \subset B_n$ and the extreme cases are $B_0(\mathbf{Y}) = \{\mathbf{Y}\}$ and $B_n(\mathbf{Y}) = \mathscr{M}(n,p)$.

To define the breakdown point, we still need a distance measure $\delta(\mathbf{T},\mathbf{T}^*)$ between the observed value $\mathbf{T} = \mathbf{T}(\mathbf{Y})$ and the corrupted value $\mathbf{T}^* = \mathbf{T}(\mathbf{Y}^*)$ of the statistic. The maximum distance over all m-replacements is then

$$\delta_m(\mathbf{T};\mathbf{Y}) = \sup\{\delta(\mathbf{T}(\mathbf{Y}),\mathbf{T}(\mathbf{Y}^*)) \ : \ \mathbf{Y}^* \in B_m(\mathbf{Y})\},$$

$m = 0,...,n.$

Definition 3.6. The *breakdown point* of $\mathbf{T}$ at $\mathbf{Y}$ is then

$$BD(\mathbf{T};\mathbf{Y}) = \min\left\{\frac{m}{n} \ : \ \delta_m(\mathbf{T};\mathbf{Y}) = \infty\right\}.$$

For location statistics, a natural distance measure is $\delta(\mathbf{M},\mathbf{M}^*) = |\mathbf{M}-\mathbf{M}^*|$. For scatter matrices, one can use $\delta(\mathbf{S},\mathbf{S}^*) = \max(|\mathbf{S}^{-1}\mathbf{S}^* - \mathbf{I}|, |(S^*)^{-1}\mathbf{S} - \mathbf{I}|)$, for example. The scatter matrix becomes "useless" if one of its eigenvalues approaches 0 or ∞. Davies (1987) gave and upper bound for the breakdown points of location vectors and scatter matrices.

Theorem 3.4. *For any location statistic* $\mathbf{M}$, *for any scatter statistic* $\mathbf{S}$, *and for any (genuinely p-variate) data matrix* $\mathbf{Y}$,

$$BD(\mathbf{M};\mathbf{Y}) \leq \frac{n-p+1}{2n} \quad \text{and} \quad BD(\mathbf{S};\mathbf{Y}) \leq \frac{n-p+1}{2n}$$

The sample mean vector and sample covariance matrix have the smallest possible breakdown point $1/n$. Maronna (1976) and Huber (1981) showed that the M-statistics have relative low breakdown points, always below $1/(p+1)$. For high breakdown points, some alternative estimation techniques (e.g., S-estimates) should be used.

3.5 Influence function and asymptotics

The influence function (Hampel, 1968, 1974) is often used to measure *local robustness* of the functional or the statistic.

Definition 3.7. The *influence function* (IF) of a functional $\mathbf{T}(F)$ at F is

$$IF(\mathbf{y};\mathbf{T},F) = \lim_{\varepsilon \downarrow 0} \frac{\mathbf{T}((1-\varepsilon)F + \varepsilon \Delta_{\mathbf{y}})) - \mathbf{T}(F)}{\varepsilon}$$

(if the limit exists), where $\Delta_{\mathbf{y}}$ is the cumulative distribution function of a distribution with all probability mass at $\mathbf{y}$.

Note that the influence function may be seen as a simple derivative of a vector valued function $\varepsilon \to \mathbf{T}((1-\varepsilon)F + \varepsilon\Delta_{\mathbf{y}})$ at zero. A robust estimator should then have a bounded and continuous IF. Intuitively, in that case, a small contamination of the distribution does not have an arbitrarily large effect on the estimate. Estimates with bounded IF can be further compared using so-called *gross error sensitivity* $\sup_{\mathbf{y}} |IF(\mathbf{y};\mathbf{T},F)|$.

In the case of elliptically symmetric distributions, the influence functions of the location and scatter functionals appear to have simple expressions (Hampel et al. (1986); Croux and Haesbrock (2000)). We assume that the scatter functional S is equipped with a correction factor so that it is a consistent estimate of Σ. If $F = F_{\mu,\Sigma}$ is the cdf of $E_p(\mu,\Sigma,\rho)$ then the following holds.

Theorem 3.5. *The influence functions of location and scatter functionals,* $\mathbf{M}(F)$ *and* $\mathbf{S}(F)$*, at elliptical* $F = F_{\mu,\Sigma}$ *are*

$$IF(\mathbf{y};\mathbf{M},F) = \gamma(r)\Sigma^{1/2}\mathbf{u}$$

and

$$IF(\mathbf{y};\mathbf{S},F) = \alpha(r)\Sigma^{1/2}\mathbf{u}\mathbf{u}'\Sigma^{1/2} - \beta(r)\Sigma,$$

where $r = |\mathbf{z}|$ *and* $\mathbf{u} = |\mathbf{z}|^{-1}\mathbf{z}$ *with* $\mathbf{z} = \Sigma^{-1/2}(\mathbf{y}-\mu)$*, and* γ*,* α*, and* β *are real valued functions determined by* $\mathbf{M}$ *and* $\mathbf{S}$ *and the spherical distribution* $F_{\mathbf{0},\mathbf{I}}$.

If $\mathbf{T}(\mathbf{Y}) = \mathbf{T}(F_n)$ is the sample version of the functional $\mathbf{T}(F)$ and the functional is sufficiently regular, then often (and this should of course be proven separately for each statistic)

$$\sqrt{n}(\mathbf{T}(\mathbf{Y})-\mathbf{T}(F)) = \sqrt{n}\,\mathbf{AVE}\{IF(\mathbf{y}_i;\mathbf{T},F)\}+o_P(1).$$

See Huber (1981). Then, under general conditions, using the central limit theorem and the influence functions given in Theorem 3.5,

$$\sqrt{n}(\mathbf{M}(\mathbf{Y})-\mu) \to_D N_p(\mathbf{0},\sigma^2\Sigma)$$

and the limiting distribution of $\sqrt{n}\,(\mathrm{vec}(\mathbf{S}(\mathbf{Y}))-\mathrm{vec}(\Sigma))$ is

$$N_p\left(\mathbf{0},\tau_1(I_{p,p}+K_{p,p})(\Sigma\otimes\Sigma)+\tau_3\mathrm{vec}(\Sigma)(\mathrm{vec}(\Sigma))'\right),$$

where now

$$\sigma^2 = \frac{E[\gamma^2(r_i)]}{p},$$
$$\tau_1 = \frac{E[\alpha^2(r_i)]}{p(p+2)}, \quad \text{and}$$
$$\tau_3 = \frac{E[\alpha^2(r_i)]}{p(p+2)} - \frac{2E[\alpha(r_i)\beta(r_i)]}{k} + E[\beta^2(r_i)]$$

with $r_i = |\mathbf{z}_i|$ with $\mathbf{z}_i = \Sigma^{-1/2}(\mathbf{y}_i-\mu)$. Note that constants σ^2, τ_1, and τ_3 are related but not the same as the finite sample constants discussed in Section 3.3.

The influence functions of the M-functionals may be found in Huber (1981). It is then interesting to note that

$$\gamma(r) \propto w_1(r)r,$$
$$\alpha(r) \propto w_2(r)r^2 \quad \text{and}$$
$$\beta(r)+\text{constant} \propto w_2(r)r^2.$$

Thus in M-estimation the weight functions determine the local robustness and efficiency properties of $\mathbf{M}$ and $\mathbf{S}$. Recall that the *Huber's M-statistic*, for example, is obtained with choices

$$w_1(r)=\min(c/r,1) \quad \text{and} \quad w_2(r)=d\cdot\min(c^2/r^2,1)$$

which then guarantee the boundedness of the influence function.

3.6 Other uses of location and scatter statistics

The scatter matrices $\mathbf{S}(\mathbf{Y})$ are often used to transform the dataset. If one writes (spectral or eigenvalue decomposition)

$$\mathbf{S}(\mathbf{Y})=\mathbf{O}(\mathbf{Y})\mathbf{D}(\mathbf{Y})(\mathbf{O}(\mathbf{Y}))'$$

where $\mathbf{O}(\mathbf{Y})$ is an orthogonal matrix and $\mathbf{D}(\mathbf{Y})$ is a diagonal matrix with positive diagonal elements in a decreasing order, then the components of the transformed data matrix

$$\mathbf{Z} = \mathbf{Y}\mathbf{O}(\mathbf{Y}),$$

are the so called principal components, used in *principal component analysis (PCA)*. The columns of $\mathbf{O}$ are also called *eigenvectors* of $\mathbf{S}$, and the diagonal elements of $\mathbf{D}$ list the corresponding *eigenvalues*. The principal components are uncorrelated and ordered according to their dispersion in the sense that $\mathbf{S}(\mathbf{Z}) = \mathbf{D}$. Principal components are often used to reduce the dimension of the data. The idea is then to take just a few first principal components combining most of the variation; the remaining components are thought to represent the noise.

Scatter matrices are also often used to standardize the data. The transformed standardized dataset

$$\mathbf{Z} = \mathbf{Y}\mathbf{O}(\mathbf{Y})[\mathbf{D}(\mathbf{Y})]^{-1/2}$$

or

$$\mathbf{Z} = \mathbf{Y}[\mathbf{S}(\mathbf{Y})]^{-1/2}$$

then has standardized components (in the sense that $\mathbf{S}(\mathbf{Z}) = \mathbf{I}_p$), and the observations $\mathbf{z}_i$ tend to be spherically distributed in the elliptic case. The symmetric version of the square root matrix

$$\mathbf{S}^{-1/2} = \mathbf{O}\mathbf{D}^{-1/2}\mathbf{O}'$$

is usually chosen. (A square root of a diagonal matrix with positive elements is a diagonal matrix of square roots of the elements.) Unfortunately, even in that case, the transformed dataset $\mathbf{Z}$ is not *coordinate-free* but the following is true.

Theorem 3.6. *For all* $\mathbf{Y}$, $\mathbf{A}$, *and* $\mathbf{S}$, *there is an orthogonal matrix* $\mathbf{O}$ *such that*

$$\mathbf{Y}\mathbf{A}'[\mathbf{S}(\mathbf{Y}\mathbf{A}')]^{-1/2}\mathbf{O} = \mathbf{Y}[\mathbf{S}(\mathbf{Y})]^{-1/2}.$$

In the independent component analysis (ICA), most ICA algorithms first standardize the data using the regular covariance matrix $\mathbf{S} = \mathbf{S}(\mathbf{Y})$ and then rotate the standardized data in such a way that the components of $\mathbf{Z} = \mathbf{Y}\mathbf{S}^{-1/2}\mathbf{O}$ are "as independent as possible". In this procedure the regular covariance matrix may be replaced by any scatter matrix that has the independence property.

A location statistic $\mathbf{M} = \mathbf{M}(\mathbf{Y})$ and a scatter matrix $\mathbf{S} = \mathbf{S}(\mathbf{Y})$ may be used together to center and standardize the dataset. Then the transformed dataset is given by

$$\mathbf{Z} = (\mathbf{Y} - \mathbf{1}_n\mathbf{M}')\mathbf{S}^{-1/2}.$$

This is often called the *whitening* of the data. Then $\mathbf{M}(\mathbf{Z}) = \mathbf{0}$ and $\mathbf{S}(\mathbf{Z}) = \mathbf{I}_p$. Again, if you rotate the dataset using an orthogonal matrix $\mathbf{O}$, it is still true that $\mathbf{M}(\mathbf{Z}\mathbf{O}) = \mathbf{0}$ and $\mathbf{S}(\mathbf{Z}\mathbf{O}) = \mathbf{I}_p$. This means that the whitening procedure is not uniquely defined. Recently, Tyler et al. (2009) developed an approach called the *invariant coordinate selection (ICS)* which is based on the simultaneous use of two scatter matrices, $\mathbf{S}_1$

and $\mathbf{S}_2$. In this procedure the data are first standardized using $\mathbf{S}_1^{-1/2}$ and then the standardized data are rotated using the principal component transformations $\mathbf{O}$ but now based on $\mathbf{S}_2$ and the transformed data $\mathbf{Y}\mathbf{S}_1^{-1/2}$. Then

$$\mathbf{S}_1(\mathbf{Y}\mathbf{S}_1^{-1/2}\mathbf{O}) = \mathbf{I}_p \quad \text{and} \quad \mathbf{S}_2(\mathbf{Y}\mathbf{S}_1^{-1/2}\mathbf{O}) = \mathbf{D},$$

where now $\mathbf{D}$ is the diagonal matrix of eigenvalues of $\mathbf{S}_1^{-1/2}\mathbf{S}_2$. If both $\mathbf{S}_1$ and $\mathbf{S}_2$ have the independence property, the procedure finds, under general assumptions, the independent components in the ICA problem.

Two location vectors and two scatter matrices may be used simultaneously to describe the skewness and kurtosis properties of a multivariate distribution. Affine invariant multivariate *skewness statistics* may be defined as squared Mahalanobis distances between two location statistics

$$|\mathbf{M}_1 - \mathbf{M}_2|_{\mathbf{S}}^2 = (\mathbf{M}_1 - \mathbf{M}_2)'\mathbf{S}^{-1}(\mathbf{M}_1 - \mathbf{M}_2).$$

The eigenvalues of $\mathbf{S}_1^{-1}\mathbf{S}_2$, say $d_1 \geq \cdots \geq d_p$ may be used to describe the *multivariate kurtosis*. The measures of skewness and kurtosis can then be used in testing for symmetry, multivariate normality, or multivariate ellipticity as well as in the separation between the models. See Kankainen et al. (2007) and Nordhausen et al. (2009b).

Chapter 4
Multivariate signs and ranks

Abstract In this chapter the concepts of multivariate spatial signs and ranks and signed-ranks are introduced. The centering and standardization of the scores are discussed. Different properties of the sign and rank scores are obtained. The sign and rank covariance matrices **UCOV**, **TCOV**, **QCOV**, and **RCOV** are introduced and discussed.

4.1 The use of score functions

Let the random sample $\mathbf{Y} = (\mathbf{y}_1, \mathbf{y}_2, ..., \mathbf{y}_n)'$ be generated by

$$\mathbf{y}_i = \mu + \Omega \varepsilon_i, \quad i = 1, ..., n,$$

where the ε_i are independent, centered, and standardized residuals with cumulative distribution function F. As discussed before, different assumptions on the distribution of ε_i yield different parametric and semiparametric models. In some applications,

$$\mathbf{Y} = \mathbf{X}\beta + \varepsilon \Omega'$$

so that the symmetry center depends on the design matrix $\mathbf{X}$. Before introducing different nonparametric scores used in our approach, we present alternative strategies on how the scores should be centered and standardized before their use in the test construction and in the estimation.

A general idea to construct tests and estimates for location parameter μ and scatter matrix $\Sigma = \Omega \Omega'$ is to use a p-vector-valued *score function* $\mathbf{T}(\mathbf{y})$ yielding individual scores $\mathbf{T}_i = \mathbf{T}(\mathbf{y}_i)$, $i = 1, ..., n$. Throughout this book we use the identity score function $\mathbf{T}(\mathbf{y}) = \mathbf{y}$, the spatial sign score function $\mathbf{U}(\mathbf{y})$, the spatial rank score function $\mathbf{R}(\mathbf{y})$, and the spatial signed-rank score function $\mathbf{Q}(\mathbf{y})$.

H. Oja, *Multivariate Nonparametric Methods with R: An Approach Based on Spatial Signs and Ranks*, Lecture Notes in Statistics 199, DOI 10.1007/978-1-4419-0468-3_4,

The general likelihood inference theory suggests that a good choice for $\mathbf{T}$ in the location problem is the optimal location score function

$$\mathbf{L}(\mathbf{y}) = \nabla \log f(\mathbf{y} - \mu)|_{\mu=\mathbf{0}},$$

that is, the gradient vector of $\log f(\mathbf{y} - \mu)$ with respect to μ at the origin. In the $N_p(\mathbf{0}, \mathbf{I}_p)$ case the optimal score function is the identity score function,

$$\mathbf{T}(\mathbf{y}) = \mathbf{y}.$$

The optimal location score function for the p-variate t-distribution with ν degrees of freedom, $t_{\nu,p}(\mathbf{0}, \mathbf{I}_p)$, for example, is

$$\mathbf{T}(\mathbf{y}) = \frac{\nu + p}{1 + |\mathbf{y}|^2}\mathbf{y},$$

and, in the general spherical model with the density function

$$f(\mathbf{y}) = \exp\{-\rho(|\mathbf{y}|)\},$$

the optimal location score function is

$$\mathbf{T}(\mathbf{y}) = \frac{\psi(|\mathbf{y}|)}{|\mathbf{y}|}\mathbf{y},$$

where $\psi(r) = \rho'(r)$, that is, the derivative function of ρ. An example of a robust choice of the score function is Huber's score function

$$\mathbf{T}(\mathbf{y}) = \min\left\{\frac{c}{|\mathbf{y}|}, 1\right\}\mathbf{y}$$

with some choice of $c > 0$. The validity, efficiency, and robustness properties of the testing and estimation procedures then naturally depend on the choice of the score function and of course on the true model.

In the test construction we first transform

$$\mathbf{Y} = (\mathbf{y}_1, ..., \mathbf{y}_n)' \rightarrow \mathbf{T} = (\mathbf{T}_1, ..., \mathbf{T}_n)' = (\mathbf{T}(\mathbf{y}_1), ..., \mathbf{T}(\mathbf{y}_n))'.$$

For different testing or estimation purposes, one then often wishes the scores to be centered and standardized in some natural way.

We then have the following possibilities.

1. *Outer centering of the scores:*

$$\mathbf{T}_i \rightarrow \hat{\mathbf{T}}_i = \mathbf{T}_i - \bar{\mathbf{T}}.$$

2. *Outer standardization of the scores:*

$$\mathbf{T}_i \rightarrow \hat{\mathbf{T}}_i = \mathbf{COV}(\mathbf{T})^{-1/2}\mathbf{T}_i.$$

3. *Outer centering and standardization of the scores:*

$$\mathbf{T}_i \rightarrow \hat{\mathbf{T}}_i = \mathbf{COV}(\mathbf{T})^{-1/2}(\mathbf{T}_i - \bar{\mathbf{T}}).$$

However, it is often more natural to use the following.

1. *Inner centering of the scores:* Find shift vector $\mathbf{M}$ such that, if $\hat{\mathbf{T}}_i = \mathbf{T}(\mathbf{y}_i - \mathbf{M})$, then
$$\mathbf{AVE}\{\hat{\mathbf{T}}_i\} = \mathbf{0}.$$
Then transform
$$\mathbf{T}_i \rightarrow \hat{\mathbf{T}}_i = \mathbf{T}(\mathbf{y}_i - \mathbf{M}).$$
2. *Inner standardization of the scores:* Find transformation matrix $\mathbf{S}^{-1/2}$ such that, if $\hat{\mathbf{T}}_i = \mathbf{T}(\mathbf{S}^{-1/2}\mathbf{y}_i)$, then
$$p \cdot \mathbf{AVE}\{\hat{\mathbf{T}}_i\hat{\mathbf{T}}_i'\} = \mathbf{AVE}\{\hat{\mathbf{T}}_i'\hat{\mathbf{T}}_i\}\mathbf{I}_p.$$
Then transform
$$\mathbf{T}_i \rightarrow \hat{\mathbf{T}}_i = \mathbf{T}(\mathbf{S}^{-1/2}\mathbf{y}_i).$$
3. *Inner centering and standardization of the scores:* Find shift vector $\mathbf{M}$ and transformation matrix $\mathbf{S}^{-1/2}$ such that, if $\hat{\mathbf{T}}_i = \mathbf{T}(\mathbf{S}^{-1/2}(\mathbf{y}_i - \mathbf{M}))$, then
$$\mathbf{AVE}\{\hat{\mathbf{T}}_i\} = \mathbf{0} \quad \text{and} \quad p \cdot \mathbf{AVE}\{\hat{\mathbf{T}}_i\hat{\mathbf{T}}_i'\} = \mathbf{AVE}\{\hat{\mathbf{T}}_i'\hat{\mathbf{T}}_i\}\mathbf{I}_p.$$
Then transform
$$\mathbf{T}_i \rightarrow \hat{\mathbf{T}}_i = \mathbf{T}(\mathbf{S}^{-1/2}(\mathbf{y}_i - \mathbf{M}))$$

Note that, in the inner approach, $\mathbf{M} = \mathbf{M}(\mathbf{Y})$ is a location statistic and $\mathbf{S} = \mathbf{S}(\mathbf{Y})$ is the scatter statistic corresponding to the score function $\mathbf{T}(\mathbf{y})$. The matrix $\mathbf{S}^{-1/2}$ is assumed to be a symmetric matrix here. Note, however, that inner centering and/or standardization may not always be possible.

In this book we mainly use the following score functions.

- *Identity score* $\mathbf{T}(\mathbf{y}) = \mathbf{y}$.
- *Spatial sign score* $\mathbf{T}(\mathbf{y}) = \mathbf{U}(\mathbf{y})$.
- *Spatial rank score* $\mathbf{T}(\mathbf{y}) = \mathbf{R}(\mathbf{y})$.
- *Spatial signed-rank score* $\mathbf{T}(\mathbf{y}) = \mathbf{Q}(\mathbf{y})$.

In different approaches we simply replace the regular identity score by nonparametric scores,

$$\mathbf{Y} = (\mathbf{y}_1, ..., \mathbf{y}_n) \quad \rightarrow \quad \mathbf{T} = (\mathbf{T}_1, ..., \mathbf{T}_n) \quad \text{or} \quad \hat{\mathbf{T}} = (\hat{\mathbf{T}}_1, ..., \hat{\mathbf{T}}_n).$$

The tests and estimates are then no longer optimal under the multivariate normality assumption but are robust and more powerful under heavy-tailed distributions. For the asymptotic behavior of the tests and estimates we need the following matrices

$$\mathbf{A} = E\left(\mathbf{T}(\varepsilon_i)\mathbf{L}(\varepsilon_i)'\right) \quad \text{and} \quad \mathbf{B} = E\left(\mathbf{T}(\varepsilon_i)\mathbf{T}(\varepsilon_i)'\right),$$

the covariance matrix between chosen score and the optimal score and the variance-covariance matrix of the chosen score. These two matrices play an important role in the following chapters.

4.2 Univariate signs and ranks

We trace the ideas from the univariate concepts of sign, rank and signed-rank.

These concepts are linked with the possibility to order the data. The ordering is done with the *univariate sign function*

$$U(y) = \begin{cases} +1, & \text{if } y > 0 \\ 0, & \text{if } y = 0 \\ -1, & \text{if } y < 0. \end{cases}$$

Consider a univariate dataset $\mathbf{Y} = (y_1, ..., y_n)'$ and assume that there are no ties. Let $-\mathbf{Y} = (-y_1, ..., -y_n)'$ be the dataset when the observations are reflected with respect to the origin. The (empirical) *centered rank function* is

$$R(y) = R_{\mathbf{Y}}(y) = \mathbf{AVE}\{U(y - y_i)\},$$

and the *signed-rank function* is

$$Q(y) = Q_{\mathbf{Y}}(y) = \frac{1}{2}[R_{\mathbf{Y}}(y) + R_{-\mathbf{Y}}(y)].$$

Note that the signed-rank function $Q_{\mathbf{Y}}(y)$ is just the centered rank function calculated among the combined $2n$-set of the original observations $y_1, ..., y_n$ and their reflections $-y_1, ..., -y_n$. The signed-rank function is odd, meaning that $Q_{\mathbf{Y}}(-y) = -Q_{\mathbf{Y}}(y)$ for all y and $\mathbf{Y}$. Note also that $R_{-\mathbf{Y}}(y) = -R_{\mathbf{Y}}(-y)$.

The numbers $U_i = U(y_i)$, $R_i = R(y_i)$, and $Q_i = Q(y_i)$, $i = 1, ..., n$, are the observed signs, observed centered ranks, and observed signed-ranks. The possible values of *observed centered ranks* R_i are

$$-\frac{n-1}{n}, -\frac{n-3}{n}, ..., \frac{n-3}{n}, \frac{n-1}{n}.$$

The possible values of *observed signed-ranks* Q_i are

$$-\frac{2n-1}{2n}, -\frac{2n-3}{2n}, ..., \frac{2n-3}{2n}, \frac{2n-1}{2n}.$$

The centered ranks and the signed-ranks are located on the interval (-1,1) (univariate unit ball). Note that the centered rank $R(x)$ and signed-rank $Q(x)$ provide both magnitude (robust distances from the median and from the origin, respectively) and direction (sign with respect to the median and sign with respect to the origin, respectively).

There are $n!$ possible values of the vector of ranks $(R_1,...,R_n)'$, given by

$$\mathbf{P}\begin{pmatrix} -(n-1)/n \\ -(n-3)/n \\ \cdots \\ (n-3)/n \\ (n-1)/n \end{pmatrix},$$

where $\mathbf{P}$ goes through all $n \times n$ permutation matrices, and if the observations are independent and identically distributed, then all these possible values have an equal probability $1/n!$. The vector of signed-ranks $(Q_1,...,Q_n)'$ has $2^n n!$ possible values obtained from

$$\mathbf{JP}\begin{pmatrix} 1/(2n) \\ 3/(2n) \\ \cdots \\ (2n-3)/(2n) \\ (2n-1)/(2n) \end{pmatrix},$$

with all possible permutation matrices $\mathbf{P}$ and all possible sign-change matrices $\mathbf{J}$, and if the distribution is symmetric around zero then all the possible values have the same probability $1/(2^n n!)$.

It is easy to find the connection between the regular rank (with values $1,2,...,n$) and the centered rank, namely,

$$\text{centered rank} = \frac{2}{n}\left[\text{regular rank} - \frac{n+1}{2}\right].$$

The above definitions of univariate signs and ranks are based on the ordering of the data. However, in the multivariate case there is no natural ordering of the data points. The approach utilizing objective or criterion functions is then needed to extend the concepts to the multivariate case. The concepts of univariate sign and rank and signed-rank may be implicitly defined using the L_1 criterion functions

$$\begin{aligned}
\mathbf{AVE}\{|y_i|\} &= \mathbf{AVE}\{U(y_i)\cdot y_i\}, \\
\frac{1}{2}\mathbf{AVE}\{|y_i-y_j|\} &= \mathbf{AVE}\{R_{\mathbf{Y}}(y_i)\cdot y_i\}, \\
\frac{1}{2}\mathbf{AVE}\{|y_i+y_j|\} &= \mathbf{AVE}\{R_{-\mathbf{Y}}(y_i)\cdot y_i\}, \quad \text{and} \\
\frac{1}{4}\mathbf{AVE}\{|y_i-y_j|+|y_i+y_j|\} &= \mathbf{AVE}\{Q_{\mathbf{Y}}(y_i)\cdot y_i\}.
\end{aligned}$$

See Hettmansperger and Aubuchon (1988).

Let us have a closer look at these objective functions if applied to the residuals in the linear regression model. We first remind the reader that the classical L_2 objective function optimal in the normal case is similarly

$$\mathbf{AVE}\{|y_i|^2\} = \mathbf{AVE}\{y_i \cdot y_i\}$$

and corresponds to the identity score function. The first objective function, the *mean deviation* of the residuals, is the basis for the so-called least absolute deviation (LAD) methods; it yields different median-type estimates and sign tests in the one-sample, two-sample, several-sample and finally general linear model settings. The second objective function is the *mean difference* of the residuals. The second, third, and fourth objective functions generate Hodges-Lehmann-type estimates and rank tests for different location problems. Note also that the sign, centered rank, and signed-rank function may be seen as the *score functions* corresponding to the first, second, and fourth objective functions. This formulation suggests a natural way to generalize the concepts sign, rank, and signed-rank to the multivariate case (without defining a multivariate ordering).

Consider next the corresponding theoretical functions, and assume that y, y_1, and y_2 are independent and identically distributed continuous random variables from a univariate distribution with cdf F. Then, similarly to empirical equations,

$$\begin{aligned}
\mathbf{E}\{|y|\} &= \mathbf{E}\{U(y)\cdot y\}, \\
\frac{1}{2}\mathbf{E}\{|y_1-y_2|\} &= \mathbf{E}\{(2F(y)-1)\cdot y\}, \\
\frac{1}{2}\mathbf{E}\{|y_1+y_2|\} &= \mathbf{E}\{(1-2F(-y))\cdot y\}, \quad \text{and} \\
\frac{1}{4}\mathbf{E}\{|y_1-y_2|+|y_1+y_2|\} &= \mathbf{E}\{U(y)[F_y(|y|)-F_y(-|y|)]\cdot y\}.
\end{aligned}$$

The theoretical (population) *centered rank function* and *signed-rank function* for F are naturally defined as

$$\begin{aligned}
R_F(y) &= 2F(y)-1 \quad \text{and} \\
Q_F(y) &= U(y)[F(|y|)-F(-|y|)].
\end{aligned}$$

If $\mathbf{Y}$ is a random sample of size n from F then it is easy to see that

$$\sup_y |R_{\mathbf{Y}}(y) - R_F(y)| \rightarrow_P 0 \quad \text{and} \quad \sup_y |Q_{\mathbf{Y}}(y) - Q_F(y)| \rightarrow_P 0$$

as $n \to \infty$. The functions $Q_{\mathbf{Y}}(y)$ and $Q_F(y)$ are odd, that is, $Q_{\mathbf{Y}}(-y) = -Q_{\mathbf{Y}}(y)$ and $Q_F(-y) = -Q_F(y)$, and for distributions symmetric about the origin $Q_F(y) = R_F(y)$. Note also that the inverse of the centered rank function (i.e., the inverse of the centered cumulative distribution function) is the univariate *quantile function*.

4.3 Multivariate spatial signs and ranks

Next we go to the multivariate case: Let $\mathbf{Y} = (\mathbf{y}_1, ..., \mathbf{y}_n)'$ be an $n \times p$ dataset. The multivariate concepts of *spatial sign*, *spatial rank* and *spatial signed-rank* are then given in the following.

Definition 4.1. The empirical *spatial sign, spatial rank, and spatial signed-rank functions* $\mathbf{U}(\mathbf{y})$, $\mathbf{R}(\mathbf{y}) = \mathbf{R}_{\mathbf{Y}}(\mathbf{y})$, and $\mathbf{Q}(\mathbf{y}) = \mathbf{Q}_{\mathbf{Y}}(\mathbf{y})$ are defined as

$$\mathbf{U}(\mathbf{y}) = \begin{cases} |\mathbf{y}|^{-1}\mathbf{y}, & \mathbf{y} \neq \mathbf{0} \\ \mathbf{0}, & \mathbf{y} = \mathbf{0} \end{cases},$$

$$\mathbf{R}(\mathbf{y}) = \mathbf{AVE}\{\mathbf{U}(\mathbf{y} - \mathbf{y}_i)\}, \quad \text{and}$$

$$\mathbf{Q}(\mathbf{y}) = \frac{1}{2}\left[\mathbf{R}_{\mathbf{Y}}(\mathbf{y}) + \mathbf{R}_{-\mathbf{Y}}(\mathbf{y})\right].$$

Observe that in the univariate case regular sign, rank, and signed-rank functions are obtained. Clearly multivariate signed-rank function $\mathbf{Q}(\mathbf{y})$ is also odd; that is, $\mathbf{Q}(-\mathbf{y}) = -\mathbf{Q}(\mathbf{y})$.

The observed *spatial signs* are $\mathbf{U}_i = \mathbf{U}(\mathbf{y}_i)$, $i = 1, ..., n$. As in the univariate case, the observed *spatial ranks* are certain averages of signs of pairwise differences

$$\mathbf{R}_i = \mathbf{R}(\mathbf{y}_i) = \mathbf{AVE}_j\{\mathbf{U}(\mathbf{y}_i - \mathbf{y}_j)\}, \quad i = 1, ..., n.$$

Finally, the observed *spatial signed-ranks* are given as

$$\mathbf{Q}_i = \mathbf{Q}(\mathbf{y}_i) = \frac{1}{2}\mathbf{AVE}_j\{\mathbf{U}(\mathbf{y}_i - \mathbf{y}_j) + \mathbf{U}(\mathbf{y}_i + \mathbf{y}_j)\}, \quad i = 1, ..., n.$$

The spatial sign $\mathbf{U}_i$ is just a direction vector of length one (lying on the unit p-sphere $\mathscr{S}_p$) whenever $\mathbf{y}_i \neq \mathbf{0}$. The centered ranks $\mathbf{R}_i$ and signed-ranks $\mathbf{Q}_i$ lie in the unit p-ball $\mathscr{B}_p$. The direction of $\mathbf{R}_i$ ($\mathbf{Q}_i$) roughly tells the direction of $\mathbf{y}_i$ from the center of the data cloud (the origin), and its length roughly tells how far away this point is from the center (the origin). The next theorem collects some equivariance properties.

Theorem 4.1. *The spatial signs, spatial ranks, and spatial signed-ranks are orthogonal equivariant in the sense that*

$$\mathbf{U}(\mathbf{O}\mathbf{y}_i) = \mathbf{O}\mathbf{U}(\mathbf{y}_i),$$
$$\mathbf{R}_{\mathbf{YO}'}(\mathbf{O}\mathbf{y}_i) = \mathbf{O}\mathbf{R}_{\mathbf{Y}}(\mathbf{y}_i), \quad \textit{and}$$
$$\mathbf{Q}_{\mathbf{YO}'}(\mathbf{O}\mathbf{y}_i) = \mathbf{O}\mathbf{Q}_{\mathbf{Y}}(\mathbf{y}_i)$$

for all $\mathbf{y}_i$ *and all orthogonal matrices* $\mathbf{O}$*. The centered ranks are invariant under location shifts and*

$$\mathrm{AVE}_i\{\mathbf{R}_i\} = \mathbf{0}.$$

Example 4.1. The spatial signs, ranks, and signed-ranks are not affine equivariant, however. In Figure 4.1 one can see scatterplots for 50 bivariate observations from $N_2(\mathbf{0}, \mathbf{I}_2)$ with the corresponding bivariate spatial signs and ranks and signed-ranks. The data points are then rescaled (Figure 4.2) and shifted (Figure 4.3). The figures illustrate the behavior of the signs, ranks, and signed-ranks under these transformations: they are not equivariant under rescaling of the components. The spatial ranks are invariant under location shifts. See below the R-code needed for the plots.

```
>library(MNM)
>set.seed(1)

>X <- rmvnorm(50,c(0,0))
>colnames(X) <- c("X_1", "X_2")

>par(mfrow = c(2, 2), pty = "s", las = 1)
>plot(X, xlim = c(-4, 4), ylim = c(-4, 4),
 main = "Original data")
>plot(spatial.sign(X, FALSE, FALSE), xlim = c(-1, 1),
 ylim = c(-1, 1),xlab = "X_1", ylab = "X_2",
 main = "Spatial signs")
>plot(spatial.rank(X, FALSE), xlim = c(-1, 1), ylim = c(-1, 1),
 xlab = "X_1", ylab = "X_2", main = "Spatial ranks")
>plot(spatial.signrank(X, FALSE, FALSE), xlim = c(-1, 1),
 ylim = c(-1, 1), , xlab = "X_1", ylab = "X_2",
 main = "Spatial signed ranks")

>X.rescaled <- transform(X, X_2 = X_2 * 5)
>plot(X.rescaled, xlim = c(-15, 15), ylim = c(-15, 15),
 main = "Rescaled data")
>plot(spatial.sign(X.rescaled, FALSE, FALSE), xlim = c(-1, 1),
 ylim = c(-1, 1), xlab = "X_1", ylab = "X_2",
 main = "Spatial signs")
>plot(spatial.rank(X.rescaled, FALSE), xlim = c(-1, 1),
 ylim = c(-1, 1), xlab = "X_1", ylab = "X_2",
 main = "Spatial ranks")
>plot(spatial.signrank(X.rescaled, FALSE, FALSE),
 xlim = c(-1, 1), ylim = c(-1, 1), , xlab = "X_1", ylab = "X_2",
```

```
 main = "Spatial signed ranks")

>X.shifted <- transform(X, X_1 = X_1 + 1)
>plot(X.shifted, xlim = c(-4, 4), ylim = c(-4, 4),
main = "Shifted data")
>plot(spatial.sign(X.shifted, FALSE, FALSE),
 xlim = c(-1, 1), ylim = c(-1, 1),
 xlab = "X_1", ylab = "X_2", main = "Spatial signs")
>plot(spatial.rank(X.shifted, FALSE), xlim = c(-1, 1),
 ylim = c(-1, 1), xlab = "X_1", ylab = "X_2",
 main = "Spatial ranks")
>plot(spatial.signrank(X.shifted, FALSE, FALSE),
 xlim = c(-1, 1), ylim = c(-1, 1), , xlab = "X_1",
 ylab = "X_2", main = "Spatial signed ranks")

>par(opar)
```

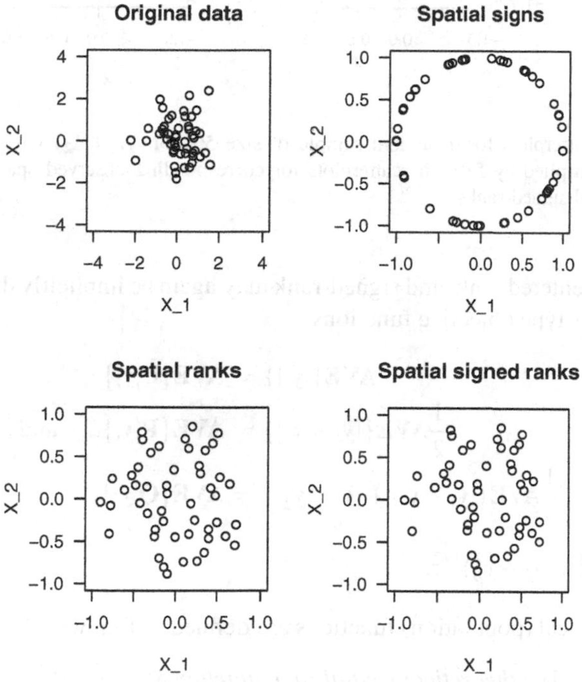

Fig. 4.1 The scatterplots for a random sample of size 50 from $N_2(\mathbf{0}, \mathbf{I}_2)$ with scatterplots for corresponding observed spatial signs, spatial ranks, and spatial signed-ranks.

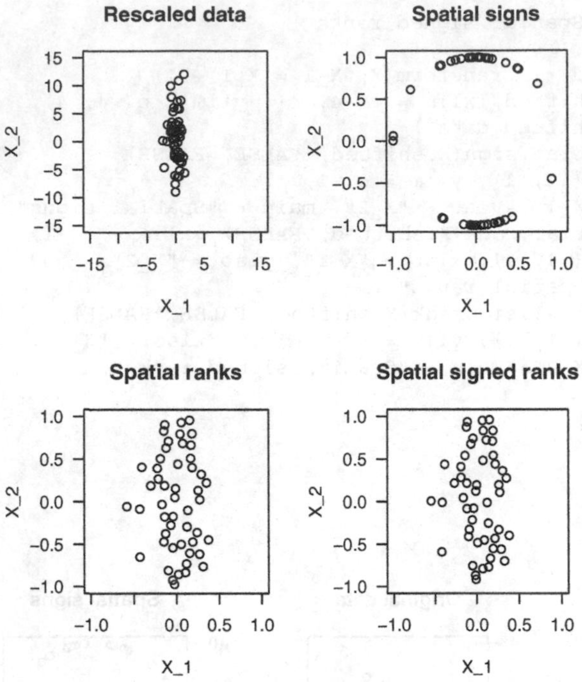

Fig. 4.2 The scatterplots for a random sample of size 50 from $N_2(\mathbf{0}, \mathbf{I}_2)$ with rescaled second component (multiplied by 5) with scatterplots for corresponding observed spatial signs, spatial ranks, and spatial signed-ranks.

The sign, centered rank, and signed-rank may again be implicitly defined through multivariate L_1 type objective functions

$$\mathbf{AVE}\{|\mathbf{y}_i|\} = \mathbf{AVE}\{\mathbf{U}_i'\mathbf{y}_i\},$$
$$\frac{1}{2}\mathbf{AVE}\{|\mathbf{y}_i - \mathbf{y}_j|\} = \mathbf{AVE}\{\mathbf{R}_i'\mathbf{y}_i\}, \quad \text{and}$$
$$\frac{1}{4}\mathbf{AVE}\{|\mathbf{y}_i - \mathbf{y}_j| + |\mathbf{y}_i + \mathbf{y}_j|\} = \mathbf{AVE}\{\mathbf{Q}_i'\mathbf{y}_i\}.$$

Here $|\mathbf{y}| = (y_1^2 + \cdots + y_p^2)^{1/2}$.

The theoretical (population) functions are defined as follows.

Definition 4.2. The *theoretical spatial rank function* and *signed-rank function* for a p-variate random variable $\mathbf{y}_i$ with cdf F are

$$\mathbf{R}_F(\mathbf{y}) = \mathbf{E}\{\mathbf{U}(\mathbf{y} - \mathbf{y}_i)\} \quad \text{and}$$
$$\mathbf{Q}_F(\mathbf{y}) = \frac{1}{2}\mathbf{E}\{\mathbf{U}(\mathbf{y} - \mathbf{y}_i) - \mathbf{U}(\mathbf{y} + \mathbf{y}_i)\}.$$

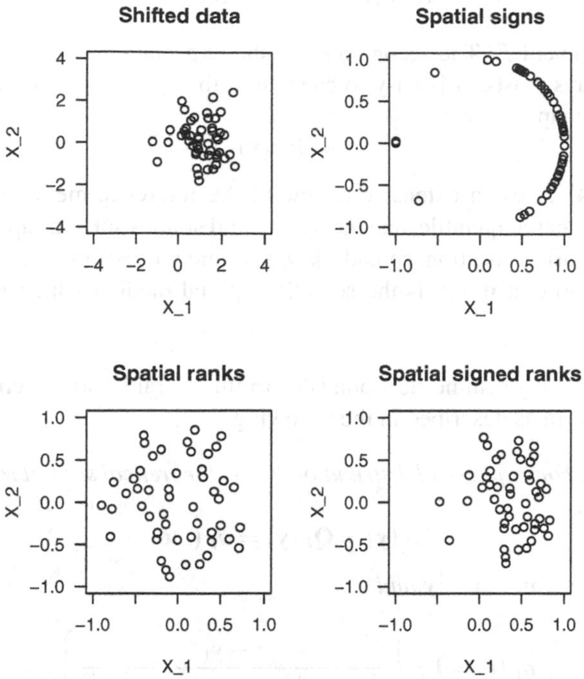

Fig. 4.3 The scatterplots for a random sample of size 50 from $N_2(\mathbf{0}, \mathbf{I}_2)$ with shifted first component (shifted by 1) with scatterplots for corresponding observed spatial signs, spatial ranks, and spatial signed-ranks.

The rank function $\mathbf{R}_F(\mathbf{y})$ characterizes the distribution F (up to a location shift). If we know the rank function, we know the distribution (up to a location shift). See Koltchinskii (1997). For F symmetric around the origin, $\mathbf{Q}_F(\mathbf{y}) = \mathbf{R}_F(\mathbf{y})$, for all $\mathbf{y}$. The empirical functions converge uniformly in probability to the theoretical ones under mild assumptions. (For the proof, see Möttönen et al. (1997).)

Theorem 4.2. *Assume that* $\mathbf{Y}$ *is a random sample of size* n *from a distribution with cdf* F *and uniformly bounded density. Then, as* $n \to \infty$,

$$\sup_{\mathbf{y}} |\mathbf{R}_{\mathbf{Y}}(\mathbf{y}) - \mathbf{R}_F(\mathbf{y})| \to_P 0 \quad \textit{and}$$

$$\sup_{\mathbf{y}} |\mathbf{Q}_{\mathbf{Y}}(\mathbf{y}) - \mathbf{Q}_F(\mathbf{y})| \to_P 0.$$

Chaudhuri (1996) considered the inverse of the spatial rank function and called it the *spatial quantile function*. See also Koltchinskii (1997). Let $\mathbf{u}$ be a vector in the p-variate open unit ball $\mathscr{B}_p$. Write $\Phi(\mathbf{u}, \mathbf{y}) = |\mathbf{y}| - \mathbf{u}'\mathbf{y}$. Then, according to Chaudhuri's definition, the value of spatial quantile function $\theta = \theta(\mathbf{u})$ of F at $\mathbf{u}$ minimizes

$$\mathbf{E}\{\Phi(\mathbf{u},\theta-\mathbf{y})-\Phi(\mathbf{u},\mathbf{y})\},$$

where $\mathbf{y}$ has the cdf F. The second term in the expectation guarantees that the expectation always exists. It is easy to check that the spatial quantile function is the inverse of the map

$$\mathbf{y} \to \mathbf{R}_F(\mathbf{y}).$$

Serfling (2004) gives an extensive review of the inference methods based on the concept of the spatial quantile, and introduces and studies some nonparametric measures of multivariate location, spread, skewness and kurtosis in terms of these quantiles. The quantile at $\mathbf{u}=\mathbf{0}$ is the so-called spatial median which is discussed in more detail later.

If F is spherically symmetric around the origin, the rank and signed-rank function has a simple form as described in the following.

Theorem 4.3. *For a spherical distribution F, the theoretical spatial rank and signed-rank function is*

$$\mathbf{R}_F(\mathbf{y})=\mathbf{Q}_F(\mathbf{y})=q_F(r)\mathbf{u},$$

where $r=|\mathbf{y}|$ and $\mathbf{u}=|\mathbf{y}|^{-1}\mathbf{y}$ and

$$q_F(r)=\mathbf{E}_F\left\{\frac{r-y_1}{((r-y_1)^2+y_2^2+\cdots+y_p^2)^{1/2}}\right\}.$$

Note that $q_F(r)$ is the derivative function of

$$D_F(r)=\mathbf{E}_F(|r\mathbf{e}_1-\mathbf{y}|-|\mathbf{y}|).$$

(The expected value always exists.) Naturally also $|q_F(r)|\le 1$, so $\mathbf{R}_F(\mathbf{y})$ is in the unit ball.

In the following examples we give general formulas for the spatial rank function in cases of multivariate normal, multivariate t, and multivariate normal scale mixture models. For these distributions, the spatial rank functions can be formulated with the generalized hypergeometric functions.

Definition 4.3. A generalized hypergeometric function is defined as a series

$${}_pF_q(a_1,a_2,\ldots,a_p;b_1,b_2,\ldots,b_q;r)=\sum_{i=0}^{\infty}\frac{(a_1)_i(a_2)_i\ldots(a_p)_i}{(b_1)_i(b_2)_i\ldots(b_q)_i}\frac{r^i}{i!},$$

where $(c)_i=\Gamma(c+i)/\Gamma(c)$.

It is straightforward but tedious then to get the theoretical rank functions in the following cases. See Möttönen et al. (2005).

Example 4.2. In the case of a multivariate normal distribution $N_p(\mathbf{0},\mathbf{I}_p)$ we get

$$D_0(r) = 2^{1/2}\exp\left\{-\frac{r^2}{2}\right\}\frac{\Gamma(\frac{p+1}{2})}{\Gamma(\frac{p}{2})}\,{}_1F_1\left(\frac{p+1}{2};\frac{p}{2};\frac{r^2}{2}\right)$$

and

$$q_0(r) = \frac{r}{2^{1/2}}\exp\left\{-\frac{r^2}{2}\right\}\frac{\Gamma(\frac{p+1}{2})}{\Gamma(\frac{p+2}{2})}\,{}_1F_1\left(\frac{p+1}{2};\frac{p+2}{2};\frac{r^2}{2}\right).$$

Example 4.3. Let $\phi(\mathbf{y};\mu,\Sigma)$ be the density of a multivariate normal distribution with mean vector μ and the covariance matrix Σ. Then a standardized multivariate normal scale mixture density is given by

$$\int \phi(\mathbf{y};\mathbf{0},s^{-2}\mathbf{I}_p)dH(s).$$

Under general assumptions, one then easily gets

$$D(r) = \int D_0(rs)dH(s) \quad \text{and} \quad q(r) = \int q_0(rs)dH(s).$$

Example 4.4. Multivariate $t_{p,\nu}$ distribution is obtained if $\nu s^2 \sim \chi^2(\nu)$. Then

$$D(r) = \frac{\nu^{1/2}\Gamma(\frac{p+1}{2})\Gamma(\frac{\nu-1}{2})}{\Gamma(\frac{\nu}{2})\Gamma(\frac{p}{2})[1+r^2/\nu]^{(\nu-1)/2}}\,{}_2F_1\left(\frac{p+1}{2},\frac{\nu-1}{2};\frac{p}{2};\frac{r^2/\nu}{1+r^2/\nu}\right)$$

and

$$q(r) = \frac{r\Gamma(\frac{p+1}{2})\Gamma(\frac{\nu+1}{2})}{\nu^{1/2}\Gamma(\frac{\nu}{2})\Gamma(\frac{p+2}{2})[1+r^2/\nu]^{(\nu+1)/2}}\,{}_2F_1\left(\frac{p+1}{2},\frac{\nu+1}{2};\frac{p+2}{2};\frac{r^2/\nu}{1+r^2/\nu}\right).$$

Example 4.5. In the case of the mixture of two normal distributions, $N_p(\mathbf{0},\mathbf{I}_p)$ with the probability π_1 and $N_p(\mathbf{0},\sigma^2\mathbf{I}_p)$ with the probability π_2, $\pi_1+\pi_2=1$, then

$$D(r) = \pi_1 D_0(r) + \pi_2 D_0(r/\sigma) \quad \text{and} \quad q(r) = \pi_1 q_0(r) + \pi_2 q_0(r/\sigma).$$

4.4 Sign and rank covariance matrices

If spatial sign and ranks are used to analyze the data, the *sign and rank covariance matrices* also naturally play an important role.

Definition 4.4. Let $\mathbf{Y} \in \mathscr{M}(n,p)$ be a data matrix. Then the *spatial sign covariance matrix* $\mathbf{UCOV(Y)}$, and *the spatial Kendall's tau matrix* $\mathbf{TCOV(Y)}$ are

$$\mathbf{UCOV(Y)} = \mathbf{AVE}\{\mathbf{U}(\mathbf{y}_i)\mathbf{U}(\mathbf{y}_i)'\},$$
$$\mathbf{TCOV(Y)} = \mathbf{AVE}\{\mathbf{U}(\mathbf{y}_i-\mathbf{y}_j)\mathbf{U}(\mathbf{y}_i-\mathbf{y}_j)'\}.$$

We also define the following.

Definition 4.5. Let $\mathbf{Y} \in \mathscr{M}(n,p)$ be a data matrix. Then the *spatial rank covariance matrix* $\mathbf{RCOV}(\mathbf{Y})$ and the *spatial signed-rank covariance matrix* $\mathbf{QCOV}(\mathbf{Y})$ are

$$\mathbf{RCOV}(\mathbf{Y}) = \mathbf{AVE}\{\mathbf{R}_i\mathbf{R}_i'\} \quad \text{and}$$
$$\mathbf{QCOV}(\mathbf{Y}) = \mathbf{AVE}\{\mathbf{Q}_i\mathbf{Q}_i'\}.$$

Note that if one uses the vectors of marginal signs and ranks instead of the spatial signs and ranks, then the regular sign covariance matrix (**UCOV**), Kendall's tau matrix (**TCOV**) and Spearman's rho matrix (**RCOV**) are obtained.

The matrices $\mathbf{UCOV}(\mathbf{Y})$, $\mathbf{RCOV}(\mathbf{Y})$, $\mathbf{TCOV}(\mathbf{Y})$, and $\mathbf{QCOV}(\mathbf{Y})$ are not scatter matrices as they are not affine equivariant. They are equivariant under orthogonal transformations only. The **RCOV** and **TCOV** are shift invariant. Note also that the sign covariance matrix and the Kendall's tau matrix are standardized in the sense that $tr(\mathbf{UCOV}(\mathbf{Y})) = tr(\mathbf{TCOV}(\mathbf{Y})) = 1$.

The *theoretical (population) spatial sign and Kendall's tau covariance matrices* are defined in the following.

Definition 4.6. Let $\mathbf{y}$, $\mathbf{y}_1$, and $\mathbf{y}_2$ be independent observations from a p-variate distribution with cdf F. Then the theoretical *spatial sign covariance matrix* $\mathbf{UCOV}(F)$, and *spatial Kendall's tau matrix* $\mathbf{TCOV}(F)$ are

$$\mathbf{UCOV}(F) = \mathbf{E}\{\mathbf{U}(\mathbf{y})\mathbf{U}(\mathbf{y})'\}, \quad \text{and}$$
$$\mathbf{TCOV}(F) = \mathbf{E}\{\mathbf{U}(\mathbf{y}_1-\mathbf{y}_2)\mathbf{U}(\mathbf{y}_1-\mathbf{y}_2)'\}.$$

Visuri et al. (2000) proved the following interesting result.

Theorem 4.4. *Let F be the cumulative distribution of an elliptical random variable* $\mathbf{y}$ *centered around the origin. Let the eigenvalues of* $\mathbf{COV}(F)$ *be distinct. Then the eigenvalues of* $\mathbf{UCOV}(F)$ *and* $\mathbf{TCOV}(F)$ *are also distinct, and the eigenvectors of* $\mathbf{COV}(F)$, $\mathbf{UCOV}(F)$, *and* $\mathbf{TCOV}(F)$ *(in the decreasing order of the corresponding eigenvalues) are the same.*

The result suggests that, in the elliptic model, the spatial sign and rank covariance matrices can be used in the principal component analysis (PCA) to find the principal components. Let $\mathbf{D}$, $\mathbf{D_U}$, and $\mathbf{D_T}$ be the (diagonal) matrices of distinct eigenvalues of $\mathbf{COV}(F)$, $\mathbf{UCOV}(F)$, and $\mathbf{TCOV}(F)$, respectively, in a decreasing order. Then

$$\mathbf{D_U} = \mathbf{D_T} = \mathbf{E}\left(\frac{\mathbf{D}^{1/2}\mathbf{u}\mathbf{u}'\mathbf{D}^{1/2}}{\mathbf{u}'\mathbf{D}\mathbf{u}}\right),$$

where $\mathbf{u}$ is uniformly distributed on the unit sphere. See Locantore et al. (1999), Marden (1999b), Visuri et al. (2000, 2003), and Croux et al. (2002) for these robust alternatives to classical PCA.. For more discussion on PCA based on spatial signs and ranks, see Chapter 9.

The theoretical spatial rank and signed-rank covariance matrices are given in the following.

Definition 4.7. Let $\mathbf{y}$ be a p-variate random variable with cumulative distribution function F, spatial rank function $\mathbf{R}_F(\mathbf{y})$ and spatial signed-rank function $\mathbf{Q}_F(\mathbf{y})$. Then the theoretical *spatial rank covariance matrix* $\mathbf{RCOV}(F)$, and *spatial signed-rank covariance matrix* $\mathbf{QCOV}(F)$ are

$$\mathbf{RCOV}(F) = E\left\{\mathbf{R}_F(\mathbf{y})\mathbf{R}_F(\mathbf{y})'\right\} \quad \text{and}$$
$$\mathbf{QCOV}(F) = E\left\{\mathbf{Q}_F(\mathbf{y})\mathbf{Q}_F(\mathbf{y})'\right\}.$$

Using Corollary 3.2, the moments of the sign covariance matrix are easily found in the spherically symmetric case.

Theorem 4.5. *Assume that* $\mathbf{Y}$ *is a random sample from a spherical distribution around the origin. Then* $\mathbf{UCOV}(\mathbf{Y})$ *is distribution-free (its distribution does not depend on the distribution of the modulus), and*

$$E(\mathbf{UCOV}(\mathbf{Y})) = \frac{1}{p}\mathbf{I}_p \quad \textit{and} \quad \mathbf{COV}(vec(\mathbf{UCOV}(\mathbf{Y}))) = \frac{2}{np(p+2)}(\mathbf{P}_1 + \mathbf{P}_2),$$

where as before

$$\mathbf{P}_1 + \mathbf{P}_2 = \frac{1}{2}(\mathbf{I}_{p,p} + \mathbf{K}_{p,p}) - \frac{1}{p}\mathbf{J}_{p,p}.$$

Proof. Let $\mathbf{y}_1, ..., \mathbf{y}_n$ be a random sample from a spherical distribution, and let $\mathbf{U}_i = |\mathbf{y}_i|^{-1}\mathbf{y}_i$, $i = 1, ..., n$. Then

$$\mathbf{UCOV}(\mathbf{Z}) = \mathbf{AVE}\{\mathbf{U}_i\mathbf{U}_i'\}.$$

Clearly $p \cdot \mathbf{E}[\mathbf{UCOV}(\mathbf{Z})] = \mathbf{I}_p$. For the variances and covariances, recall from Theorem 2.2 that $(U_{i1}^2, ..., U_{ip}^2)$ has a Dirichlet distribution $D_p(1/2, ..., 1/2)$, so

$$\mathbf{E}[U_{ij}^2] = \frac{1}{p}, \quad \mathbf{E}[U_{ij}^2 U_{ik}^2] = \frac{1}{p(p+2)} \quad \text{and} \quad \mathbf{E}[U_{ij}^4] = \frac{3}{p(p+2)},$$

with $j \neq k$. But then

$$Var(U_{ij}^2) = \frac{2}{p(p+2)}, \quad Var(U_{ij}U_{ik}) = \frac{1}{p(p+2)} \quad \text{and} \quad Cov(U_{ij}^2, U_{ik}^2) = -\frac{2}{p^2(p+2)},$$

with $j \neq k$, and the result follows from Corollary 3.2.

Rank covariance matrices $\mathbf{TCOV}(\mathbf{Y})$, $\mathbf{RCOV}(\mathbf{Y})$, and $\mathbf{QCOV}(\mathbf{Y})$ are (asymptotically equivalent to) matrix-valued U-statistics with kernel sizes 2, 3, and 3, respectively; they are not (even asymptotically) distribution-free, and the expressions for their covariance matrices are more complicated: For spherical distributions around the origin, the first and second moments of $\mathbf{TCOV}(\mathbf{Y})$ have a similar structure as those of the $\mathbf{UCOV}(\mathbf{Y})$ with expected value $(1/p)\mathbf{I}_p$ and covariance matrix

$$\tau(\mathbf{P}_1+\mathbf{P}_2)$$

but τ now depends on sample size n and the distribution of the modulus. The limiting distributions of $\mathbf{UCOV}(\mathbf{Y})$ and $\mathbf{TCOV}(\mathbf{Y})$ can be easily given in the spherical case. See Sirkiä et al. (2008).

Theorem 4.6. *Assume that* $\mathbf{Y}$ *is a random sample of size* n *from a spherical distribution around the origin. Then, as* $n \to \infty$,

$$\sqrt{n}vec\left(\mathbf{UCOV}(\mathbf{Y})-\frac{1}{p}\mathbf{I}\right) \to_d N_{p^2}\left(\mathbf{0},\frac{2}{p(p+2)}(\mathbf{P}_1+\mathbf{P}_2)\right) \quad and$$
$$\sqrt{n}vec\left(\mathbf{TCOV}(\mathbf{Y})-\frac{1}{p}\mathbf{I}\right) \to_d N_{p^2}\left(\mathbf{0},\frac{2\tau}{p(p+2)}(\mathbf{P}_1+\mathbf{P}_2)\right),$$

where

$$\tau = 4\mathbf{E}\left[\left(1-\frac{|\mathbf{y}_1|}{|\mathbf{y}_1-\mathbf{y}_2|}\right)\left(1-\frac{|\mathbf{y}_1|}{|\mathbf{y}_1-\mathbf{y}_3|}\right)\right]$$

as $n \to \infty$.

The expected values of $\mathbf{RCOV}(\mathbf{Y})$ and $\mathbf{QCOV}(\mathbf{Y})$ when $\mathbf{Y}$ is a random sample from F satisfy

$$\mathbf{E}(\mathbf{RCOV}(\mathbf{Y})) = \mathbf{RCOV}(F)+o(1/n) \quad \text{and}$$
$$\mathbf{E}(\mathbf{QCOV}(\mathbf{Y})) = \mathbf{QCOV}(F)+o(1/n)$$

so that the sample statistics are asymptotically unbiased. They estimate the same population quantity if F is symmetrical around the origin. In the spherically symmetric case their covariance matrices are structured as

$$\tau_2(\mathbf{P}_1+\mathbf{P}_2)+\tau_3\mathbf{P}_3.$$

See Corollaries 3.1 and 3.2.

4.5 Other approaches

Next we give a short survey of some other approaches to multivariate sign and rank methods. The most straightforward extension is just to use *vectors of componentwise univariate signs and ranks*: the componentwise signs and ranks then correspond to the L_1 type criterion functions

$$\mathbf{AVE}_i\{|y_{i1}|+\cdots+|y_{ip}|\} \quad \text{and} \quad \mathbf{AVE}_{i,j}\{|y_{i1}-y_{j1}|+\cdots+|y_{ip}-y_{jp}|\}$$

utilizing the so called "Manhattan distance". The sign and rank vectors are invariant under componentwise monotone transformations (e.g., marginal rescaling) but not orthogonal equivariant. This approach is perhaps the most natural one for data with independent components. See Puri and Sen (1971) for a complete discussion of this approach.

Affine equivariant multivariate sign and rank methods may be constructed as follows. The geometric idea in the p-variate location model then is to consider data based simplices based on $p+1$ residuals, or p residuals and the origin. The volume of the simplex with vertices $\mathbf{y}_{i_1},...,\mathbf{y}_{i_{p+1}}$ is

$$V(\mathbf{y}_{i_1},...,\mathbf{y}_{i_{p+1}}) = \frac{1}{p!}\text{abs}\left\{\det\begin{pmatrix} 1 & \cdots & 1 \\ \mathbf{y}_{i_1} & \cdots & \mathbf{y}_{i_{p+1}} \end{pmatrix}\right\}.$$

The two criterion functions now are

$$\mathbf{AVE}\{V(\mathbf{0},\mathbf{y}_{i_1},...,\mathbf{y}_{i_p})\} \quad \text{and} \quad \mathbf{AVE}\left\{V(\mathbf{y}_{i_1},...,\mathbf{y}_{i_{p+1}})\right\}.$$

The corresponding score functions are then *affine equivariant or Oja multivariate signs and ranks*, and the inference methods based on these are then affine equivariant/invariant. For a review of the multivariate location problem, see Oja (1999). Visuri et al. (2000, 2003) and Ollila et al. (2003b, 2004) introduced and considered the corresponding affine equivariant sign and rank covariance matrices.

Koshevoy and Mosler (1997a,b, 1998) and Mosler (2002) proposed the use of *zonotopes* and *lift zonotopes*, p- and $(p+1)$-variate convex sets $Z_p(\mathbf{Y})$ and $LZ_{p+1}(\mathbf{Y})$, respectively, to describe and investigate the properties of a data matrix $\mathbf{Y}$. Koshevoy et al. (2003) developed a scatter matrix estimate based on the zonotopes. It appears that there is a nice duality relation between zonotopes (lift zonotopes) and affine equivariant signs (ranks); the objective functions yielding affine equivariant signs and ranks are just volumes of zonotope $Z_p(\mathbf{Y})$ and lift zonotope $LZ_{p+1}(\mathbf{Y})$, respectively. See Koshevoy et al. (2004).

Randles (1989) developed an affine invariant sign test based on an ingenious concept of *interdirection counts*. Affine invariant interdirection counts depend on the directions of the observation vectors; they measure the angular distances between two vectors relative to the rest of the data. Randles (1989) was followed by a series of papers introducing nonparametric sign and rank interdirection tests for multivariate one sample and two sample location problems, for example. This approach is quite related to the spatial sign and rank approach, as we show in later chapters.

Still one important approach is to combine the directions (spatial signs or interdirection counts) and the transformed ranks of the Mahalanobis distances from the origin or data center. In a series of papers, Hallin and Paindaveine constructed *optimal signed-rank location tests* in the elliptical model; see the seminal papers by

Hallin and Paindaveine (2002, 2006). Similarly, Nordhausen et al. (2009) developed optimal rank tests in the independent component model.

Chapter 5
One-sample problem: Hotelling's T^2-test

Abstract We start with a one-sample location example with trivariate and bivariate observations. It is shown how a general score function $\mathbf{T}(\mathbf{y})$ is used to construct tests and estimates in the one-sample location problem. The identity score $\mathbf{T}(\mathbf{y})$ gives the regular Hotelling's T^2-test and sample mean.

5.1 Example

We consider the classical data due to Rao (1948) consisting of weights of cork borings on trees in four directions: north (N), east (E), south (S), and west (W). We have these four measurements on 28 trees, and we wish to test whether the weight of cork borings is independent of the direction.

Table 5.1 Weights of cork borings (in centigrams) in the four directions

N	E	S	W	N	E	S	W
72	66	76	77	91	79	100	75
60	53	66	63	56	68	47	50
56	57	64	58	79	65	70	61
41	29	36	38	81	80	68	58
32	32	35	36	78	55	67	60
30	35	34	26	46	38	37	38
39	39	31	27	39	35	34	37
42	43	31	25	32	30	30	32
37	40	31	25	60	50	67	54
33	29	27	36	35	37	48	39
32	30	34	28	39	36	39	31
63	45	74	63	50	34	37	40
54	46	60	52	43	37	39	50
47	51	52	43	48	54	57	43

H. Oja, *Multivariate Nonparametric Methods with R: An Approach Based on Spatial Signs and Ranks*, Lecture Notes in Statistics 199, DOI 10.1007/978-1-4419-0468-3_5,

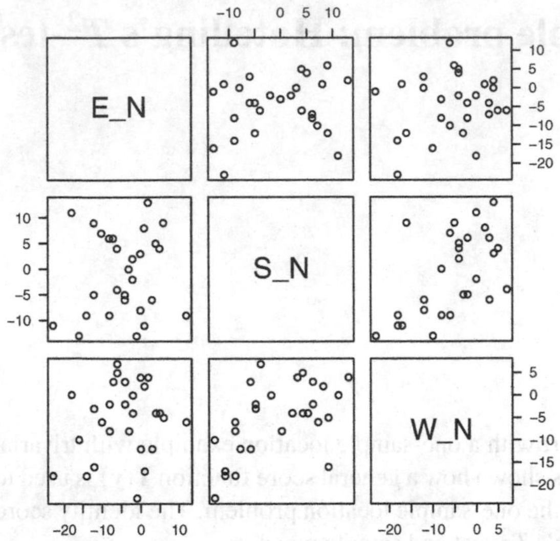

Fig. 5.1 The scatterplot for differences E-N, S-N and W-N.

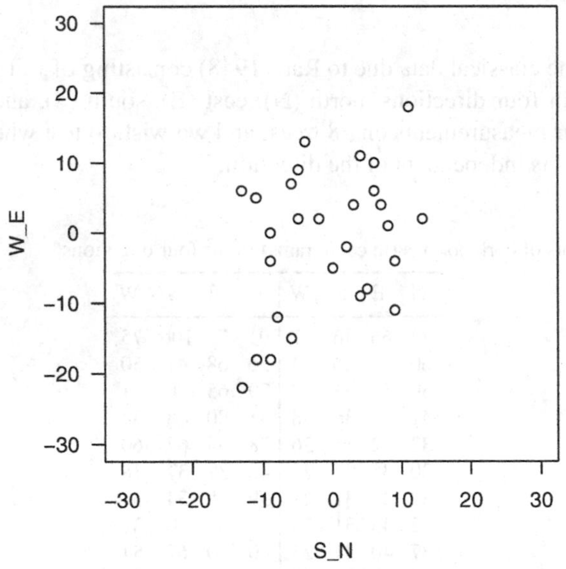

Fig. 5.2 The scatterplot for differences S-N and W-E.

```
>data(cork)

>cork_3v <- sweep(cork[,2:4], 1, cork[,1], "-")
>colnames(cork_3v) <- c("E_N", "S_N", "W_N")
>pairs(cork_3v, las = 1)


>cork_2v <- with(cork,
 data.frame(S_N = South - North, W_E = West - East))
>plot(cork_2v, xlim = c(-30, 30), ylim = c(-30, 30),
 las = 1, pty = "s")
```

The independence of the weight of cork borings on the direction is considered with a 3-variate variable consisting on differences E-N, S-N and W-N when north (N) is subtracted from the other three. See Figure 5.1. Also, the symmetry of cork boring is studied with a bivariate vector of differences S-N and W-E. See Figure 5.2. We wish to test the null hypothesis that the distributions of (E-N,S-N,W-N) as well as that of (S-N,W-E) are symmetric around zero.

5.2 General strategy for estimation and testing

Let the random sample $\mathbf{Y} = (\mathbf{y}_1, \mathbf{y}_2, ..., \mathbf{y}_n)'$ be generated by

$$\mathbf{y}_i = \mu + \Omega \varepsilon_i, \quad i = 1, ..., n,$$

where the ε_i are centered and standardized residuals with cumulative distribution function F. The tests and estimates are constructed under different symmetry assumptions (A0)–(A4) and (B0)–(B4). Note that the zero vector may be used as a null value without loss of generality, because to test $H_0: \ \mu = \mu_0$, we just substitute $\mathbf{y}_i - \mu_0$ in place of $\mathbf{y}_i$ in the tests.

We now describe the use of a general score function $\mathbf{T}(\mathbf{y})$ for the statistical inference in the one sample location problem. The results are only heuristic and general, and the distributional assumptions for the asymptotic theory of course depend on the chosen score function. For the one-sample symmetry center problem it is natural to assume that the score function $\mathbf{T}(\mathbf{y})$ is odd; that is, $\mathbf{T}(-\mathbf{y}) = -\mathbf{T}(\mathbf{y})$ for all $\mathbf{y}$.

Outer standardization. We first discuss the test and estimate that use the outer standardization. The test may not be affine invariant, and the estimate may not be affine equivariant. Now

- *The test statistic is*

$$\mathbf{T}(\mathbf{Y}) = \mathbf{AVE}\{\mathbf{T}_i\} = \mathbf{AVE}\{\mathbf{T}(\mathbf{y}_i)\}.$$

- *The companion location estimate $\hat{\mu}$ is the shift vector obtained in the inner centering, and is determined by estimating equations*

$$\mathbf{AVE}\{\mathbf{T}(\mathbf{y}_i - \hat{\mu})\} = \mathbf{0}.$$

For the asymptotic theory in our approach we need the following $p \times p$ matrices $\mathbf{A}$ and $\mathbf{B}$ (expectations taken under the null hypothesis),

$$\mathbf{A} = \mathbf{E}\{\mathbf{T}(\mathbf{y}_i)\mathbf{L}(\mathbf{y}_i)'\} \quad \text{and} \quad \mathbf{B} = \mathbf{E}\left\{\mathbf{T}(\mathbf{y}_i)\mathbf{T}(\mathbf{y}_i)'\right\}.$$

Consider the null hypothesis $H_0: \ \mu = \mathbf{0}$ and a (contiguous) sequence of alternatives $H_n: \ \mu = n^{-1/2}\delta$. The alternatives are used to consider the asymptotic relative efficiencies (ARE) of the tests and estimates. They can also be used in sample size calculations. Then, under general assumptions, we get the following results.

- *Under the null hypothesis H_0,*

$$\sqrt{n}\,\mathbf{AVE}\{\mathbf{T}_i\} \to_d N_p(\mathbf{0}, \mathbf{B}).$$

- *Under the null hypothesis H_0, the squared version of the test statistic*

$$Q^2 = n\,|\hat{\mathbf{B}}^{-1/2}\mathbf{AVE}\{\mathbf{T}_i\}|^2 \to_d \chi^2_p,$$

where

$$\hat{\mathbf{B}} = \mathbf{AVE}\left\{\mathbf{T}_i\mathbf{T}_i'\right\}.$$

- *Under the sequence of alternatives H_n,*

$$\sqrt{n}\,\mathbf{AVE}\{\mathbf{T}_i\} \to_d N_p(\mathbf{A}\delta, \mathbf{B}).$$

- *The limiting distribution of the estimate $\hat{\mu}$ is given by*

$$\sqrt{n}(\hat{\mu} - \mu) \to_d N_p(\mathbf{0}, \mathbf{A}^{-1}\mathbf{B}\mathbf{A}^{-1}).$$

Inner standardization. Inner standardization is sometimes used in the construction of the test and the estimate. If the inner standardization is possible, then the test is affine invariant, and the estimate is affine equivariant.

- *In the inner standardization of the test, one first finds a $p \times p$ transformation matrix $\mathbf{S}^{-1/2}$ such that, for $\hat{\mathbf{T}}_i = \mathbf{T}(\mathbf{S}^{-1/2}\mathbf{y}_i)$, $i = 1, ..., n$,*

$$p \cdot \mathbf{AVE}\left\{\hat{\mathbf{T}}_i\hat{\mathbf{T}}_i'\right\} = \mathbf{AVE}\left\{\hat{\mathbf{T}}_i'\hat{\mathbf{T}}_i\right\}\mathbf{I}_p.$$

The squared form of the test statistic is then

$$Q^2 = np \cdot \frac{|\mathbf{AVE}\{\hat{\mathbf{T}}_i\}|^2}{\mathbf{AVE}\{|\hat{\mathbf{T}}_i|^2\}}$$

with a limiting chi-square distribution with p degrees of freedom.

- *Find a shift vector $\hat{\mu}$ and a transformation matrix $\mathbf{S}^{-1/2}$ such that, for $\hat{\mathbf{T}}_i = \mathbf{T}(\mathbf{S}^{-1/2}(\mathbf{y}_i - \hat{\mu}))$, $i = 1,...,n$,*

$$\mathbf{AVE}\{\hat{\mathbf{T}}_i\} = \mathbf{0} \quad and \quad p \cdot \mathbf{AVE}\{\hat{\mathbf{T}}_i\hat{\mathbf{T}}_i'\} = \mathbf{AVE}\{\hat{\mathbf{T}}_i'\hat{\mathbf{T}}_i\}\mathbf{I}_p.$$

Then $\hat{\mu}$ is the location estimate based on inner standardization.

Note that, in the testing case, $\mathbf{S}$ is not a regular scatter matrix estimate but a scatter matrix with respect to a known center (the origin). In the estimation case, $\mathbf{S}$ is a regular scatter matrix (around the estimated value $\hat{\mu}$).

For later extensions to the several-sample and regression cases we next give the test statistics in a slightly different form. The test statistics is then seen to compare two different scatter matrices. For that purpose, write

$$\mathbf{P_X} = \mathbf{X}(\mathbf{X}'\mathbf{X})^{-1}\mathbf{X}'$$

for any $n \times q$ matrix $\mathbf{X}$ with rank $q < n$. Matrix $\mathbf{P_X}$ is the $p \times p$ projection matrix to the subspace spanned by the columns of $\mathbf{X}$. The transformation $\mathbf{Y} \to \mathbf{P}_{\mathbf{1}_n}\mathbf{Y}$ then just replaces all the observations by their sample mean vector. Now, in outer standardization,

$$\mathbf{Y} \quad \to \quad \mathbf{T} \quad \to \quad Q^2 = n \cdot tr((\mathbf{T}'\mathbf{P}_{\mathbf{1}_n}\mathbf{T})(\mathbf{T}'\mathbf{T})^{-1}),$$

and, in inner standardization,

$$\mathbf{Y} \quad \to \quad \hat{\mathbf{T}} \quad \to \quad Q^2 = n \cdot tr((\hat{\mathbf{T}}\mathbf{P}_{\mathbf{1}_n}\hat{\mathbf{T}})(\hat{\mathbf{T}}'\hat{\mathbf{T}})^{-1}).$$

The test statistic based on inner standardization (if possible) is affine invariant. This is not necessarily true if one uses outer standardization.

The approximate p-value may thus be based on the limiting chi-square distribution. For small sample sizes, an alternative way to construct the p-value is to use the *sign-change argument*. Let $\mathbf{J}$ be an $n \times n$ diagonal matrix with diagonal elements ± 1. It is called a sign-change matrix. The value of the test statistic for a sign-changed sample $\mathbf{JY}$ is then

$$Q^2(\mathbf{JY}) = n \cdot tr((\mathbf{T}'\mathbf{J}\mathbf{P}_{\mathbf{1}_n}\mathbf{J}\mathbf{T})(\mathbf{T}'\mathbf{T})^{-1}) \quad \text{or} \quad n \cdot tr((\hat{\mathbf{T}}\mathbf{J}\mathbf{P}_{\mathbf{1}_n}\mathbf{J}\hat{\mathbf{T}})(\hat{\mathbf{T}}'\hat{\mathbf{T}})^{-1})$$

($\mathbf{T}'\mathbf{T}$ and $\hat{\mathbf{T}}'\hat{\mathbf{T}}$ are invariant under sign changes). Then the p-value of a conditionally distribution-free sign-change test statistic is

$$\mathbf{E_J}\{I(Q^2(\mathbf{JY}) \geq Q^2(\mathbf{Y}))\},$$

where $\mathbf{J}$ has a uniform distribution over its all 2^n possible values. This sign-change version of the test is valid for the null hypothesis $-\mathbf{y} \sim \mathbf{y}$ (model (A4)).

5.3 Hotelling's T^2-test

Let $\mathbf{Y} = (\mathbf{y}_1, ..., \mathbf{y}_n)'$ be a random sample from an unknown distribution, and assume that the p-variate observation vectors $\mathbf{y}_i$ are generated by

$$\mathbf{y}_i = \mu + \Omega \varepsilon_i, \quad i = 1, ..., n,$$

where the ε_i are centered and standardized random vectors with the cumulative distribution function F. In the Hotelling's test case, we assume that the standardized vectors ε_i have mean vector zero and covariance matrix $\mathbf{I}_p$. Then $\mathbf{E}(\mathbf{y}_i) = \mu$ is an unknown mean vector, and $\mathbf{COV}(\mathbf{y}_i) = \Sigma = \Omega\Omega' > 0$ an unknown covariance matrix. At first, we wish to estimate unknown μ and test the null hypothesis

$$H_0: \ \mu = \mathbf{0},$$

and we use the identity score function $\mathbf{T}(\mathbf{y}) = \mathbf{y}$. Then we get

Hotelling's T^2 and the sample mean: Hotelling's test is obtained with score function $\mathbf{T}(\mathbf{y}) = \mathbf{y}$. Then

$$\mathbf{T}(\mathbf{Y}) = \bar{\mathbf{y}} \quad \textit{and also} \quad \hat{\mu} = \bar{\mathbf{y}}$$

and, under the null hypothesis,

$$\mathbf{A} = \mathbf{I}_p \quad \textit{and} \quad \mathbf{B} = \mathbf{C} = \Sigma.$$

Both the outer and inner standardizations yield the same Hotelling's one-sample test statistic

$$Q^2 = Q^2(\mathbf{Y}) = n\bar{\mathbf{y}}'\hat{\mathbf{B}}^{-1}\bar{\mathbf{y}}.$$

In the above test construction, we standardized the sample mean using the sample covariance matrix with respect to the origin, $\mathbf{B}(\mathbf{Y}) = \mathbf{AVE}\{\mathbf{y}_i\mathbf{y}_i'\}$. The popular version of Hotelling's T^2 uses the regular sample mean and regular sample covariance matrix, namely $\bar{\mathbf{y}}$ and

$$\mathbf{C} = \mathbf{AVE}\{(\mathbf{y}_i - \bar{\mathbf{y}})(\mathbf{y}_i - \bar{\mathbf{y}})'\},$$

and is defined as

$$T^2 = T^2(\mathbf{Y}) = n\bar{\mathbf{y}}'\mathbf{C}^{-1}\bar{\mathbf{y}}.$$

Under the null hypothesis, both $\mathbf{B}(\mathbf{Y})$ and

$$\mathbf{C}(\mathbf{Y}) = \mathbf{B}(\mathbf{Y}) - \bar{\mathbf{y}}\bar{\mathbf{y}}'$$

converge in probability to the true value Σ.

In the testing procedure with sample size n, instead of reporting a p-value, one often compares the observed value of the test statistic Q_n^2 to a critical value c_n. The null hypothesis is rejected if $Q_n^2 > c_n$. The validity and asymptotic validity of a test (Q_n^2, c_n) for the null hypothesis H_0 is defined as follows.

Definition 5.1. The test (Q_n^2, c_n) is a *valid level-α test* for sample size n if $P_{H_0}(Q_n^2 > c_n) = \alpha$. The sequence of tests (Q_n^2, c_n), $n = 1, 2, \ldots$, is *asymptotically valid* with level α if $P_{H_0}(Q_n^2 > c_n) \to \alpha$. The probabilities P_{H_0} are calculated under the null hypothesis H_0.

Model of multivariate normality We now recall some well-known properties of Hotelling's T^2-test in the case where the observations come from a multivariate normal distribution $N_p(\mu, \Sigma)$. (The distribution of ε_i is $N(\mathbf{0}, \mathbf{I}_p)$.) The Hotelling's T^2 test rejects H_0 if T^2 is large enough. Hotelling's T^2 is a valid test statistic in this model as its exact distribution under H_0 is known. The large values of T^2 support the alternative $\mu \neq \mathbf{0}$. The distribution of

$$\frac{n-p}{np} T^2$$

is known to be an F-distribution with p and $n-p$ degrees of freedom, denoted by $F_{p,n-p}$. The exact p-value is then obtained as the tail probability of $F_{p,n-p}$. Under the alternative ($\mu \neq \mathbf{0}$), the distribution of T^2 is known to be a noncentral $F_{p,n-p}$ with the noncentrality parameter $n\mu'\Sigma^{-1}\mu$; one can therefore calculate the exact power under a fixed alternative as well.

Nonparametric model with bounded second moments. Hotelling's test statistic (both versions Q^2 and T^2) can be used in a larger model where we only assume that the ε_i have mean vector zero and covariance matrix identity. Note that the distribution does not have to be symmetric; only the assumption on the existence of the second moments is needed. Under this wider model, the test statistic T^2 is still asymptotically valid as its limiting distribution under H_0 is known. Again, large values of T^2 support the alternative $\mu \neq \mathbf{0}$. If the second moments exist ($E(|\varepsilon_i \varepsilon_i'|) < \infty$), then the central limit theorem (CLT) can be used to show that the limiting distribution of Q^2 (and T^2) is a chi-square distribution with p degrees of freedom. The test is *asymptotically unbiased* as, under any fixed alternative $\mu \neq \mathbf{0}$ and any $c > 0$, $P(Q^2 > c) \to 1$, as $n \to \infty$. Under the sequence of alternative hypotheses $H_n:\ \mu = n^{-1/2}\delta$, the limiting distribution of $n^{-1/2}\bar{\mathbf{y}}$ is $N_p(\delta, \Sigma)$, and, consequently, the limiting distribution of $Q^2 = Q^2(\mathbf{Y})$ is a noncentral chi-square distribution with p degrees of freedom and noncentrality parameter $\delta'\Sigma^{-1}\delta$. This result can be used to calculate approximate power as, for μ close to zero, Q^2 has an approximate noncentral chi square distribution with p degrees of freedom and noncentrality parameter

$n\mu'\Sigma^{-1}\mu$. Also the sample size calculations to attain fixed size and power may be based on this result.

Nonparametric model with symmetry assumption. The sign-change version of Hotelling's T^2 test can be used under the symmetry assumption

$$\mathbf{J}(\varepsilon_1,\ldots,\varepsilon_n)' \sim (\varepsilon_1,\ldots,\varepsilon_n)',$$

where $\mathbf{J}$ is a $n \times n$ sign-change matrix. Note that no assumption on the existence of the moments is needed here. Next note that $\mathbf{B}(\mathbf{JY}) = \mathbf{B}(\mathbf{Y})$ but $\mathbf{C}(\mathbf{JY}) = \mathbf{C}(\mathbf{Y})$ is not necessarily true. This is to the advantage of our version Q^2 as one then simply has

$$Q^2(\mathbf{Y}) = \mathbf{1}_n'\mathbf{Y}(\mathbf{Y}'\mathbf{Y})^{-1}\mathbf{Y}\mathbf{1}_n \quad \text{and} \quad Q^2(\mathbf{JY}) = \mathbf{1}_n'\mathbf{JY}(\mathbf{Y}'\mathbf{Y})^{-1}\mathbf{YJ}\mathbf{1}_n$$

and the p-value obtained from the sign-change or exact version of the test

$$\mathbf{E}_\mathbf{J}\left\{I\left(Q^2(\mathbf{JY}) \geq Q^2(\mathbf{Y})\right)\right\}$$

is more easily calculated.

Test statistic Q^2 (as well as T^2) is *affine invariant* in the sense that

$$Q^2(\mathbf{YH}') = Q^2(\mathbf{Y})$$

for any full-rank $\mathbf{H}$. This implies that the null distribution does not depend on Ω at all. If we write

$$\hat{\varepsilon} = \mathbf{Y}\hat{\mathbf{B}}^{-1/2},$$

for the estimated residuals, then

$$Q^2(\mathbf{Y}) \;=\; Q^2(\hat{\varepsilon}) \;=\; n\cdot|\mathbf{AVE}\{\hat{\varepsilon}_i\}|^2.$$

Note that the mean-squared length of the standardized observations is $\mathbf{AVE}\{|\hat{\varepsilon}_i|^2\} = p$.

In the multivariate normality case, the sample mean $\bar{\mathbf{y}}$ and the (slightly adjusted) sample covariance matrix $n/(n-1)\mathbf{C}(\mathbf{Y})$ are optimal (uniformly minimum variance unbiased, UMVU) estimators of unknown μ and Σ. Also, it is known that

$$\bar{\mathbf{y}} \sim N_p\left(\mu, \frac{1}{n}\Sigma\right),$$

and a natural estimate of the covariance matrix of $\hat{\mu}$ is

$$\widehat{\mathbf{COV}}(\hat{\mu}) = \frac{1}{n}\mathbf{C}(\mathbf{Y}).$$

In the wider model where we only assume that

$$\mathbf{E}(\varepsilon_i) = \mathbf{0} \quad \text{and} \quad \mathbf{COV}(\varepsilon_i) = \mathbf{I}_p,$$

the limiting distribution of $\sqrt{n}(\bar{\mathbf{y}} - \mu)$ is a multivariate normal distribution $N_p(\mathbf{0}, \Sigma)$ as well, and therefore the distribution of $\bar{\mathbf{y}}$ may be approximated by $N_p(\mu, (1/n)\,\mathbf{C}(\mathbf{Y}))$. The estimated covariance matrix is thus $(1/n)\mathbf{C}(\mathbf{Y})$ and the approximate 95% confidence ellipsoid is given by

$$\left\{\mu \,:\, n(\mu - \bar{\mathbf{y}})\mathbf{C}^{-1}(\mu - \bar{\mathbf{y}}) \leq \chi^2_{p,0.95}\right\}.$$

Example 5.1. **Cork boring data** Consider first the 3-variate vector of E-N, S-N and W-N. If we wish to test the null hypothesis that the mean vector is zero, we get

```
> mv.1sample.test(X1)

        Hotelling's one sample T2-test

data:  X
T.2 = 20.742, df = 3, p-value = 0.0001191
alternative hypothesis: true location is not equal to c(0,0,0)
```

If we then wish to estimate the population mean vector with the sample mean vector, the estimate and its estimated covariance matrix are as given below. The estimate and its estimated covariance matrix are then used to find the 95% confidence ellipsoid for the population mean vector. This is given in Figure 5.4.

```
> est <- mv.1sample.est(cork_3v)
> summary(est)
The sample mean vector of cork_3v is:
     E_N     S_N     W_N
-4.3571 -0.8571 -5.3571

And has the covariance matrix:
       E_N    S_N    W_N
E_N 2.2440 0.2598 0.4199
S_N 0.2598 2.2691 1.2585
W_N 0.4199 1.2585 2.2810
>
> plotMvloc(est, X=cork_3v, color.ell=c(1,1,1))
>
```

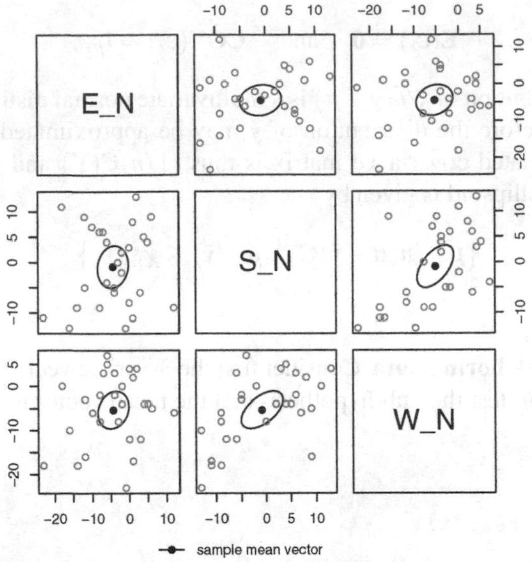

Fig. 5.3 The scatterplot with estimated mean and 95% confidence ellipsoid for 3-variate data.

Example 5.2. Consider then the bivariate vector of S-N and W-E. If we wish to test the null hypothesis that the mean vector is zero, we get

```
> mv.1sample.test(cork_2v)

        Hotelling's one sample T2-test

data:  X
T.2 = 0.4433, df = 2, p-value = 0.8012
alternative hypothesis: true location is not equal to c(0,0)
```

Again, the mean vector, its estimated covariance matrix, and 95 % confidence ellipsoids are given by

```
> est <- mv.1sample.est(cork_2v)
> summary(est)
The sample mean vector of cork_2v is:
    S_N     W_E
-0.8571 -1.0000
```

```
And has the covariance matrix:
       S_N    W_E
S_N 2.2691 0.9987
W_E 0.9987 3.6852
> plotMvloc(est, X=cork_2v, color.ell=c(1,1,1))
>
```

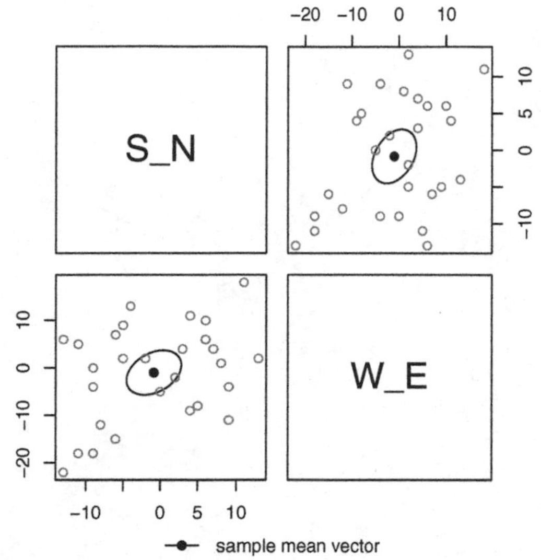

Fig. 5.4 The scatterplot with estimated mean and 95% confidence ellipse for 2-variate data.

Chapter 6
One-sample problem: Spatial sign test and spatial median

Abstract The spatial sign score function $\mathbf{U}(\mathbf{y})$ is used for the one-sample location problem. The test is then the spatial sign test, and the estimate is the spatial median. The tests and estimates using outer standardization as well as those using inner standardization are discussed.

6.1 Multivariate spatial sign test

6.1.1 Preliminaries

The aim is to find, in the one-sample location problem, test statistics that are valid under much weaker conditions than Hotelling's T^2. We consider a multivariate generalization of the univariate sign test, perhaps the simplest test ever proposed. The spatial sign test uses the spatial sign score $\mathbf{U}(\mathbf{y})$ which is given by

$$\mathbf{U}(\mathbf{y}) = |\mathbf{y}|^{-1}\mathbf{y}, \quad \text{for } \mathbf{y} \neq \mathbf{0},$$

and $\mathbf{U}(\mathbf{0}) = \mathbf{0}$.

We start by giving some approximations and key results needed in the following. Let $\mathbf{y} \neq \mathbf{0}$ and μ be any p-vectors, $p > 1$, and write

$$r = |\mathbf{y}| \quad \text{and} \quad \mathbf{u} = |\mathbf{y}|^{-1}\mathbf{y}.$$

H. Oja, *Multivariate Nonparametric Methods with R: An Approach Based on Spatial Signs and Ranks*, Lecture Notes in Statistics 199, DOI 10.1007/978-1-4419-0468-3_6,

The accuracies of different (constant, linear and quadratic) approximations of the function $|\mathbf{y}-\mu|$ of μ are given by the following.

Lemma 6.1.

1.

$$||\mathbf{y}-\mu|-|\mathbf{y}|| \le |\mu|.$$

2.

$$||\mathbf{y}-\mu|-|\mathbf{y}|-\mathbf{u}'\mu| \le 2\frac{|\mu|^2}{r}.$$

3.

$$\left||\mathbf{y}-\mu|-|\mathbf{y}|-\mathbf{u}'\mu-\mu'\frac{1}{2r}[\mathbf{I}_p-\mathbf{u}\mathbf{u}']\mu\right| \le C\frac{|\mu|^{2+\delta}}{r^{1+\delta}}$$

for all $0<\delta<1$ *where* C *does not depend on* $\mathbf{y}$ *or* μ.

In a similar way, the accuracies of constant and linear approximations of the function $|\mathbf{y}-\mu|^{-1}(\mathbf{y}-\mu)$ of μ are given by

Lemma 6.2.

1.

$$\left|\frac{\mathbf{y}-\mu}{|\mathbf{y}-\mu|}-\frac{\mathbf{y}}{|\mathbf{y}|}\right| \le 2\frac{|\mu|}{r}.$$

2.

$$\left|\frac{\mathbf{y}-\mu}{|\mathbf{y}-\mu|}-\frac{\mathbf{y}}{|\mathbf{y}|}-\frac{1}{r}[\mathbf{I}_p-\mathbf{u}\mathbf{u}']\mu\right| \le C\frac{|\mu|^{1+\delta}}{r^{1+\delta}}$$

for all $0<\delta<1$ *where* C *does not depend on* $\mathbf{y}$ *or* μ.

See Appendix B.

We also often need the following lemma.

Lemma 6.3. *Assume that the density function* $f(\varepsilon)$ *of the p-variate continuous random vector* ε *is uniformly bounded. Then* $E\{|\varepsilon|^{-\alpha}\}$ *exists for all* $0\le\alpha<2$.

6.1.2 The test outer standardization

We consider first the location model

$$\mathbf{y}_i = \mu+\varepsilon_i, \quad i=1,...,n,$$

where the independent residuals ε_i have a joint density $f(\varepsilon)$ that is uniformly bounded and they are centered so that

$$\mathbf{E}(\mathbf{U}(\varepsilon_i)) = \mathbf{0}.$$

As before, the cumulative distribution function of ε_i is denoted by $F(\varepsilon)$. The centering is needed to get an interpretation for the location parameter μ. We later show that μ is the so-called spatial median of $\mathbf{y}_i$. We wish to test the null hypothesis

$$H_0: \ \mu = \mathbf{0}.$$

The matrices

$$\mathbf{A} = \mathbf{E}\left\{|\varepsilon_i|^{-1}(\mathbf{I}_p - |\varepsilon_i|^{-2}\varepsilon_i\varepsilon_i')\right\} \quad \text{and} \quad \mathbf{B} = \mathbf{UCOV}(F) = \mathbf{E}\{|\varepsilon_i|^{-2}\varepsilon_i\varepsilon_i'\}$$

are often needed in the following. As the density function of ε_i is continuous and uniformly bounded $\mathbf{A}$ also exists and is bounded.

Multivariate spatial sign test is thus obtained with score function $\mathbf{T}(\mathbf{y}) = \mathbf{U}(\mathbf{y})$ (spatial sign score). Write $\mathbf{U}_i = \mathbf{U}(\mathbf{y}_i)$, $i = 1,...,n$. Then

$$\mathbf{T} = \mathbf{T}(\mathbf{Y}) = \mathbf{AVE}\{\mathbf{U}_i\}$$

and

$$Q^2 = Q^2(\mathbf{Y}) = n\mathbf{T}'\hat{\mathbf{B}}^{-1}\mathbf{T},$$

where

$$\hat{\mathbf{B}} = \mathbf{AVE}\{\mathbf{U}_i\mathbf{U}_i'\}.$$

Note that $\mathbf{T}(\mathbf{Y})$ is of course not a location statistic; it is only orthogonal equivariant ($\mathbf{T}(\mathbf{YO}') = \mathbf{OT}(\mathbf{Y})$).

Totally nonparametric model. The sign test statistic is asymptotically valid under extremely weak assumptions. For testing $H_0 : \mu = \mathbf{0}$ we only have to assume that the observations $\mathbf{y}_i$ are centered around μ in the sense that

$$E\{\mathbf{U}(\mathbf{y}_i - \mu)\} = \mathbf{0}.$$

Note that this assumption is naturally true under symmetry, that is, if $-(\mathbf{y}_i - \mu) \sim (\mathbf{y}_i - \mu)$. As $\mathbf{B} = \mathbf{E}(\mathbf{U}_i\mathbf{U}_i')$ always exists, the weak law of large numbers (WLLN, Kolmogorov) implies that

$$\hat{\mathbf{B}} \to_P \mathbf{B}.$$

Then, using the central limit theorem (Lindeberg-Lévy) and Slutsky's lemma, we easily obtain the following.

Theorem 6.1. *If the null hypothesis $H_0 : \ \mu = \mathbf{0}$ is true then $\sqrt{n}\mathbf{T} \to_d N(\mathbf{0}, \mathbf{B})$ and the test statistic with outer standardization*

$$Q^2 = n\mathbf{T}'\hat{\mathbf{B}}^{-1}\mathbf{T} \to \chi^2_p.$$

It is remarkable that no assumptions are made on the distribution of the modulus $|\mathbf{y}_i - \mu|$. The modulus may even depend on direction $\mathbf{U}_i = |\mathbf{y}_i - \mu|^{-1}(\mathbf{y}_i - \mu)$, and so even skewed distributions are allowed. No assumptions on moments are needed.

The finite-sample power of the test can be approximated using the following theorem.

Theorem 6.2. *Assume that* $\mathbf{E}\{\mathbf{U}(\mathbf{y}_i - \mu)\} = \mathbf{0}$ *and that the density of* ε_i *is uniformly bounded. Then, under the sequence of alternative distributions* $H_n: \ \mu = n^{-1/2}\delta$, *the limiting distribution of* $\sqrt{n}\mathbf{T}(\mathbf{Y})$ *is* $N_p(\mathbf{A}\delta, \mathbf{B})$, *and*

$$Q^2(\mathbf{Y}) \to \chi^2_p(\delta'\mathbf{A}\mathbf{B}^{-1}\mathbf{A}\delta),$$

a noncentral chi-square distribution with p degrees of freedom and noncentrality parameter $\delta'\mathbf{A}\mathbf{B}^{-1}\mathbf{A}\delta$.

Proof. Lemmas 6.2 and 6.3 together imply that

$$\begin{aligned}\sqrt{n}\mathbf{AVE}\left\{\mathbf{U}\left(\mathbf{y}_i + \frac{\delta}{\sqrt{n}}\right)\right\} &= \sqrt{n}\mathbf{AVE}\{\mathbf{U}(\mathbf{y}_i)\} \\ &+ \mathbf{AVE}\left\{\frac{1}{|\mathbf{y}_i|}(\mathbf{I}_p - \mathbf{U}(\mathbf{y}_i)\mathbf{U}(\mathbf{y}_i)')\right\}\delta + o_P(1).\end{aligned}$$

See also Möttönen et al. (1997).

Nonparametric model with symmetry assumption. An exact *sign-change version* of the test is obtained if we can assume that, under the null hypothesis,

$$\mathbf{JU} \sim \mathbf{U}$$

for all $n \times n$ sign-change matrices $\mathbf{J}$. Here $\mathbf{U} = (\mathbf{U}_1, ..., \mathbf{U}_n)'$. Note that sign covariance matrix $\mathbf{UCOV}(\mathbf{Y}) = \mathbf{UCOV}(\mathbf{JY})$, for all $\mathbf{J}$ (invariance under sign changes). Therefore

$$Q^2(\mathbf{Y}) = \mathbf{1}_n'\mathbf{U}(\mathbf{U}'\mathbf{U})^{-1}\mathbf{U}'\mathbf{1}_n \quad \text{and} \quad Q^2(\mathbf{JY}) = \mathbf{1}_n'\mathbf{JU}(\mathbf{U}'\mathbf{U})^{-1}\mathbf{U}'\mathbf{J}\mathbf{1}_n.$$

Recall that the exact p-value is calculated as

$$\mathbf{E}\left[I\{Q^2(\mathbf{JY}) \geq Q^2(\mathbf{Y})\}\right],$$

where $I\{\cdot\}$ is an indicator function and the expected value is calculated for a uniformly distributed sign-change matrix $\mathbf{J}$ (with 2^n possible values). In practice, the expected value is naturally in often approximated by simulations from the uniform distribution of $\mathbf{J}$.

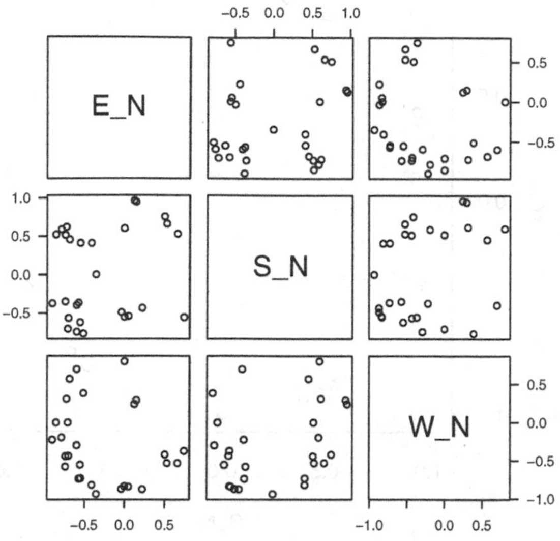

Fig. 6.1 The scatterplot for spatial signs for differences E-N, S-N and W-N. The spatial signs lie on the 3-variate unit sphere.

Example 6.1. **Cork boring data.** Consider again the 3-variate vector of E-N, S-N and W-N. The 3-variate spatial signs are illustrated in Figure 6.1. The observed value of $Q^2(\mathbf{Y})$ is 13.874 with corresponding p-value 0.003:

```
> signs_3v <- spatial.sign(cork_3v, FALSE, FALSE)
> colMeans(signs_3v)
[1] -0.28654  0.01744 -0.28271
> SCov(signs_3v, location = c(0, 0, 0))
        [,1]     [,2]     [,3]
[1,] 0.32215 0.05380 0.03225
[2,] 0.05380 0.34064 0.08021
[3,] 0.03225 0.08021 0.33721
> mv.1sample.test(cork_3v, score = "s")

        One sample spatial sign test using outer standardization

data:  cork_3v
Q.2 = 13.87, df = 3, p-value = 0.003082
alternative hypothesis: true location is not equal to c(0,0,0)

>
> pairs(signs_3v, labels = colnames(cork_3v), las = 1)
>
```

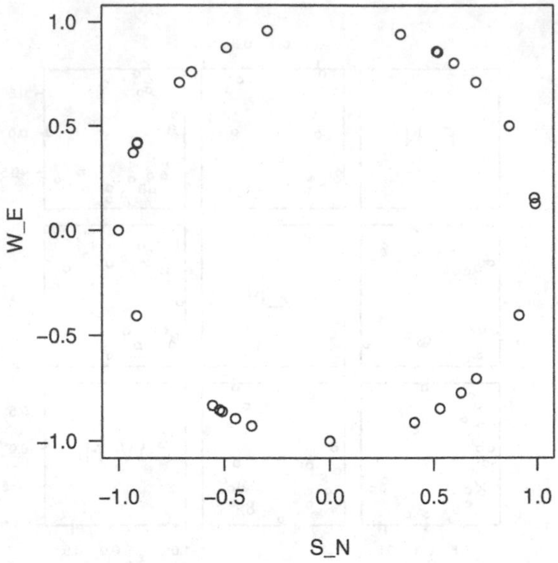

Fig. 6.2 The scatterplot for spatial signs for differences S-N and W-E.

Example 6.2. **Cork boring data.** In the bivariate case with spatial signs given in Figure 6.2. The observed value of $Q^2(\mathbf{Y})$ 0.017 with p-value 0.991. With the R-package,

```
> signs_2v <- spatial.sign(cork_2v, FALSE, FALSE)
> colMeans(signs_2v)
[1] -0.0170622  0.0002194
> SCov(signs_2v, location = c(0, 0))
        [,1]    [,2]
[1,] 0.47083 0.01315
[2,] 0.01315 0.52917
> mv.1sample.test(cork_2v, score = "s")

        One sample spatial sign test using outer standardization

data:  cork_2v
Q.2 = 0.0173, df = 2, p-value = 0.9914
alternative hypothesis: true location is not equal to c(0,0)

>
> plot(signs_2v, xlab = "S_N", ylab = "W_E", ylim = c(-1, 1),
  xlim = c(-1, 1), las = 1, pty = "s")
>
```

6.1.3 The test with inner standardization

Despite all its nice properties listed so far, the sign test statistic with outer standardization is unfortunately not affine invariant. Then, for example, the p-value depends on the chosen coordinate system. It is, however, invariant under orthogonal transformation; that is, with outer standardization,

$$Q^2(\mathbf{YO}') = \mathbf{1}_n'\mathbf{UO}'(\mathbf{OU}'\mathbf{UO}')^{-1}\mathbf{OU}'\mathbf{1}_n = \mathbf{1}_n'\mathbf{U}(\mathbf{U}'\mathbf{U})^{-1}\mathbf{U}'\mathbf{1}_n = Q^2(\mathbf{Y}),$$

for all orthogonal $\mathbf{O}$. Theorem 3.6 then implies the following.

Theorem 6.3. *Let* $\mathbf{S} = \mathbf{S}(\mathbf{Y})$ *be any scatter matrix with respect to the origin. Then* $Q^2(\mathbf{YS}^{-1/2})$ *is affine invariant.*

How should one then choose the scatter statistic $\mathbf{S}$? It seems natural to use the scatter matrix given by inner standardization. It then appears that the resulting affine invariant sign test is distribution-free under extremely weak assumptions. Using the inner standardization we get the following.

Definition 6.1. *Tyler's transformation* $\mathbf{S}^{-1/2}$ is the transformation that makes the spatial sign covariance matrix proportional to the identity matrix,

$$p \cdot \mathbf{UCOV}(\mathbf{YS}^{-1/2}) = \mathbf{I}_p.$$

The matrix can be chosen so that $tr(\mathbf{S}) = p$; this shape matrix is then called *Tyler's scatter matrix* (with respect to the origin).

Tyler's transformation (and Tyler's shape matrix) exists under weak conditions; see Tyler (1987). Tyler's transformation tries to make the spatial signs of the transformed data points $\pm\, \mathbf{S}^{-1/2}\mathbf{y}_i$, $i = 1,...,n$, be uniformly distributed on the unit p-sphere. Tyler's shape matrix $\mathbf{S}$ and Tyler's transformation $\mathbf{S}^{-1/2}$ are surprisingly easy to compute. The iterative construction may begin with $\mathbf{S} = \mathbf{I}_p$ and an iteration step is

$$\mathbf{S} \leftarrow p\, \mathbf{S}^{1/2}\mathbf{UCOV}(\mathbf{YS}^{-1/2})\, \mathbf{S}^{1/2}.$$

If $|p\ \mathbf{UCOV}(\mathbf{YS}^{-1/2}) - \mathbf{I}_p|$ is sufficiently small, then stop and fix the scale by $\mathbf{S} \leftarrow [p/tr(\mathbf{S})]\mathbf{S}$. Tyler (1987) gives weak conditions under which the algorithm converges.

Tyler's shape matrix $\mathbf{S}(\mathbf{Y})$ is thus calculated with respect to the origin. It is interesting to note that

1. Its value depends on $\mathbf{y}_i$ only through $\mathbf{U}_i = |\mathbf{y}_i|^{-1}\mathbf{y}_i$.
2. It is affine equivariant in the sense that

$$\mathbf{S}(\mathbf{YH}') \propto \mathbf{HS}(\mathbf{Y})\mathbf{H}',$$

for all datasets $\mathbf{Y}$ and all full rank matrices $\mathbf{H}$.

Consider in the following the location-scatter model

$$\mathbf{y}_i = \mu + \Omega\varepsilon_i, \quad i = 1,...,n,$$

where the independent residuals ε_i have a uniformly bounded density and the ε_i are centered and standardized so that

$$\mathbf{E}(\mathbf{U}(\varepsilon_i)) = \mathbf{0} \quad \text{and} \quad p \cdot \mathbf{E}(\mathbf{U}(\varepsilon_i)\mathbf{U}(\varepsilon_i)') = \mathbf{I}_p$$

(true in model (B2)). Then under the null hypothesis Tyler's shape matrix $\mathbf{S}(\mathbf{Y})$ converges in probability to $[p/tr(\Sigma)]\Sigma$ where $\Sigma = \Omega\Omega'$.

The spatial signs of the Tyler transformed observations, $\hat{\mathbf{U}}_i = \mathbf{U}(\mathbf{S}^{-1/2}\mathbf{y}_i)$, $i = 1,...,n$, are called *standardized spatial signs*. As before, we then write

$$\hat{\mathbf{U}} = (\hat{\mathbf{U}}_1,...,\hat{\mathbf{U}}_n)'$$

for the matrix of observed standardized spatial signs. The multivariate sign test based on standardized signs, the spatial sign test with inner standardization, then rejects H_0 for large values of

$$Q^2(\mathbf{Y}\mathbf{S}^{-1/2}) = \mathbf{1}_n'\hat{\mathbf{U}}(\hat{\mathbf{U}}'\hat{\mathbf{U}})^{-1}\hat{\mathbf{U}}'\mathbf{1}_n = \frac{p}{n}(\mathbf{1}_n'\hat{\mathbf{U}})^2 = np\left|\mathbf{AVE}\{\hat{\mathbf{U}}_i\}\right|^2.$$

$Q^2(\mathbf{Y}\mathbf{S}^{-1/2})$ is simply np times the *squared length of the average direction of the transformed data points*. This test was proposed and developed in Randles (2000) where the following important result is also given.

Theorem 6.4. *The spatial sign test with inner standardization, $Q^2(\mathbf{Y}\mathbf{S}^{-1/2})$, is affine invariant and strictly distribution-free in the model of (B1) of elliptical directions ($|\varepsilon_i|^{-1}\varepsilon_i$ uniformly distributed). The limiting distribution of $Q^2(\mathbf{Y}\mathbf{S}^{-1/2})$ under the null hypothesis is then a χ^2_p distribution.*

Proof. The affine invariance was proven in Theorem 6.3. The fact that the test is distribution-free under the model of elliptical directions follows from the fact that $Q^2(\mathbf{Y}\mathbf{S}^{-1/2})$ depends on the observations only through $|\varepsilon_i|^{-1}\varepsilon_i$, $i = 1,...,n$.

Assume next (without loss of generality) that $\Omega = \mathbf{I}_p$; that is, $\mathbf{Y} = \varepsilon$. Then $\mathbf{S}^{-1/2}$ is a root-n consistent estimate of $\mathbf{I}_p$ (Tyler (1987)). Thus

$$\Delta^* = \sqrt{n}(\mathbf{S}^{-1/2} - \mathbf{I}_p) = O_P(1).$$

Then write

$$\mathbf{S}^{-1/2} = \mathbf{I}_p + n^{-1/2}\Delta^*,$$

where Δ^* is thus bounded in probability. Using Lemma 6.2 we obtain

$$\frac{1}{\sqrt{n}}\sum_{i=1}^n \mathbf{U}(\mathbf{S}^{-1/2}\mathbf{y}_i) = \frac{1}{\sqrt{n}}\sum_{i=1}^n \mathbf{U}_i + \frac{1}{n}\sum_{i=1}^n (\Delta^* - \mathbf{U}_i'\Delta^*\mathbf{U}_i)\mathbf{U}_i + o_P(1).$$

For $|\Delta^*| < M$, the second term in the expansion converges uniformly in probability to zero due to its linearity with respect to the elements of Δ^* and due to the symmetry of the distribution of $\mathbf{U}_i$. Therefore

$$\frac{1}{\sqrt{n}}\sum_{i=1}^{n}\mathbf{U}(\mathbf{S}^{-1/2}\mathbf{y}_i) - \frac{1}{\sqrt{n}}\sum_{i=1}^{n}\mathbf{U}_i \to_P 0.$$

It is easy to see that both $\mathbf{B}(\mathbf{Y})$ and $\mathbf{B}(\mathbf{Y}\mathbf{S}^{-1/2})$ converge in probability to $\mathbf{B} = (1/p)\mathbf{I}_p$. Therefore $Q^2(\mathbf{Y}\mathbf{S}^{-1/2}) - Q^2(\mathbf{Y}) \to_P 0$ and the result follows.

Example 6.3. **Cork boring data.** Consider again the 3-variate vector of E-N, S-N and W-N. The 3-variate standardized spatial signs are illustrated in Figure 6.3. The observed value of Q^2 is 14.57 with corresponding p-value 0.002. Using the R package,

```
> signs_i_3v <- spatial.sign(cork_3v, FALSE, TRUE)
>
> mv.1sample.test(cork_3v, score = "s", stand = "i")

        One sample spatial sign test using inner standardization

data:   cork_3v
Q.2 = 14.57, df = 3, p-value = 0.002222
alternative hypothesis: true location is not equal to c(0,0,0)

>
> pairs(signs_i_3v, labels = colnames(cork_3v), las = 1)
>
```

Example 6.4. **Cork boring data.** In the bivariate case with standardized signs in Figure 6.4, the null hypothesis can not be rejected in this case as $Q^2(\mathbf{Y})$ is 0.012 with p-value 0.994.

```
>
> signs_i_2v <- spatial.sign(cork_2v, FALSE, TRUE)
>
> mv.1sample.test(cork_2v, score = "s", stand = "i")

        One sample spatial sign test using inner standardization

data:   cork_2v
Q.2 = 0.0117, df = 2, p-value = 0.9942
alternative hypothesis: true location is not equal to c(0,0)
```

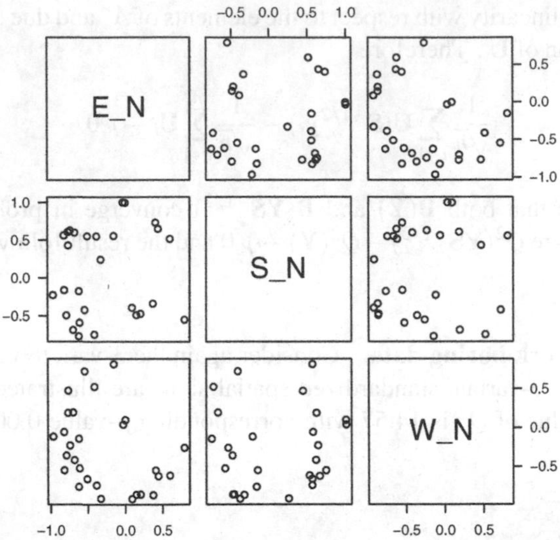

Fig. 6.3 The scatterplot for (inner) standardized spatial signs for differences E-N, S-N and W-N. The spatial signs lie in the 3-variate unit sphere.

```
>
> plot(signs_i_2v, xlab = "S_N", ylab = "W_E", ylim = c(-1, 1),
  xlim = c(-1, 1), las = 1, pty = "s")
>
```

6.1.4 Other sign-based approaches for testing problem

We show some connections to other test statistics proposed in the literature. Assume the model (B2). Then

$$E(|\varepsilon_i|^{-1}\varepsilon_i) = \mathbf{0} \quad \text{and} \quad p \cdot E(|\varepsilon_i|^{-2}\varepsilon_i\varepsilon_i') = \mathbf{I}_p.$$

If $\mu = \mathbf{0}$ and $\Omega = \mathbf{I}_p$ then

$$Q^2(\mathbf{Y}) - \frac{p}{n}\sum_{i=1}^{n}\sum_{j=1}^{n} \mathbf{U}_i'\mathbf{U}_j \to_P 0.$$

Therefore, in this case, the spatial sign test statistic is *asymptotically equivalent* to Rayleigh's statistic

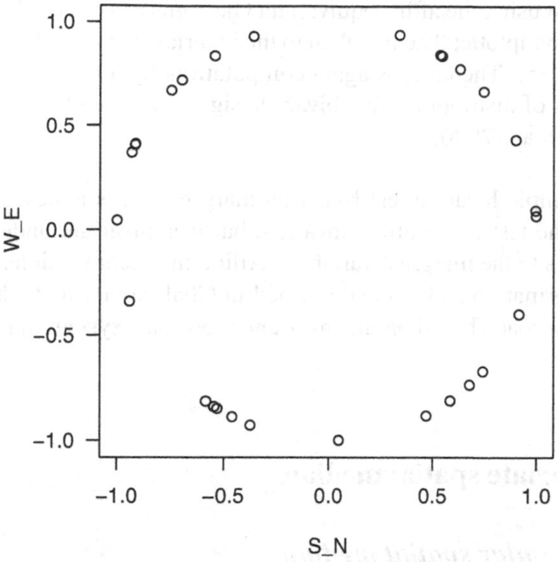

Fig. 6.4 The scatterplot for standardized spatial signs for differences S-N and W-E.

$$\frac{p}{n}\sum_{i=1}^{n}\sum_{j=1}^{n}\cos(\mathbf{U}_i,\mathbf{U}_j)=\frac{p}{n}\sum_{i=1}^{n}\sum_{j=1}^{n}\cos(\mathbf{y}_i,\mathbf{y}_j),$$

where $\cos(\mathbf{y}_i,\mathbf{y}_j)$ is the cosine of the angle between $\mathbf{y}_i$ and $\mathbf{y}_j$. Next note that if $\mathbf{S}=\mathbf{S}(\mathbf{Y})$ is Tyler's scatter matrix then

$$Q^2(\mathbf{Y}\mathbf{S}^{-1/2})=\frac{p}{n}\sum_{i=1}^{n}\sum_{j=1}^{n}\cos(\hat{\mathbf{U}}_i,\hat{\mathbf{U}}_j).$$

Randles (1989) introduced a nonparametric counterpart of $\cos(\mathbf{U}_i,\mathbf{U}_j)$ based on the so-called *interdirection counts*. His test statistic was

$$V(\mathbf{Y})=\frac{p}{n}\sum_{i=1}^{n}\sum_{j=1}^{n}\cos(\pi\hat{p}_{i,j}),$$

where the proportion $\hat{p}_{i,j}$ is the observed fraction of times that $\mathbf{y}_i$ and $\mathbf{y}_j$ fall on opposite sides of data-based hyperplanes formed by the origin and $p-1$ data points. This is an extension of the Blumen (1958) bivariate sign test. The test statistic is affine invariant and strictly distribution-free under the model (B1) of elliptical directions. It is remarkable that no scatter matrix estimate is then needed to attain affine equivariance. The test is, however, computationally difficult in high dimensions. See also Chaudhuri and Sengupta (1993) and Koshevoy et al. (2004).

The sign test using the affine equivariant Oja signs (see Oja (1999)) is also, in the elliptic case, asymptotically equivalent to the invariant version of the spatial sign test using $Q^2(\mathbf{YS}^{-1/2})$. The latter is again computationally much more convenient. For general classes of distribution-free bivariate sign tests, see Oja and Nyblom (1989) and Larocque et al. (2000).

The one-sample location test based on marginal signs is described in Puri and Sen (1971). The test is not affine invariant but it is invariant under odd monotone transformations to the marginal variables. Affine invariant versions are obtained using the transformation technique described in Chakraborty and Chaudhuri (1999). See also the approach based on the invariant coordinate system in Nordhausen et al. (2009).

6.2 Multivariate spatial median

6.2.1 The regular spatial median

Next we consider the one-sample location estimation problem. The estimate corresponding to the spatial sign test is the so-called spatial median $\hat{\mu} = \hat{\mu}(\mathbf{Y})$ which is, under general assumptions, a root-n consistent estimate of the true spatial population median μ, and has good limiting and finite-sample distributional properties. The estimate $\hat{\mu}$ is also robust with a bounded influence function and a breakdown point $1/2$. In addition, we also give an affine equivariant version of the estimate.

We introduce the spatial median using the corresponding objective function: as seen before, the spatial sign score corresponds to the sum of Euclidean distances. Recall that the sample mean vector minimizes the objective function which is the mean of squared Euclidean distances $\mathbf{AVE}\{|\mathbf{y}_i - \mu|^2\}$. A natural alternative measure of distance is just the mean of Euclidean distances, and we define the following.

Definition 6.2. The *sample spatial median* $\hat{\mu}(\mathbf{Y})$ minimizes the criterion function $\mathbf{AVE}\{|\mathbf{y}_i - \mu|\}$ or, equivalently,

$$D_n(\mu) = \mathbf{AVE}\{|\mathbf{y}_i - \mu| - |\mathbf{y}_i|\}.$$

The spatial median has a very long history starting in Weber (1909), Gini and Galvani (1929) and Haldane (1948). Gower (1974) used the term mediancenter. Brown (1983) has developed many of the properties of the spatial median. This minimization problem is also sometimes known as the Fermat-Weber location problem. Taking the gradient of the objective function, one sees that if $\hat{\mu}$ solves the equation

$$\mathbf{AVE}\{\mathbf{U}(\mathbf{y}_i - \hat{\mu})\} = \mathbf{0},$$

then $\hat{\mu}$ is the observed spatial median. This shows the connection between the spatial median and the spatial sign test. The estimate $\hat{\mu}$ is the value of the location parameter which, if used as a null value, offers the highest possible p-value. The solution can also be seen as the shift vector corresponding to the inner centering of the spatial sign score test; the location shift makes the spatial signs (directions) of the centered data points sum up to $\mathbf{0}$.

The spatial median is unique, if the dimension of the data cloud is greater than one Milasevic and Ducharme (1987). The Weiszfeld algorithm for the computation of the spatial median has a simple iteration step,

$$\mu \leftarrow \mu + \frac{\mathbf{AVE}\{\mathbf{U}(\mathbf{y}_i - \mu)\}}{\mathbf{AVE}\{|\mathbf{y}_i - \mu|^{-1}\}}.$$

The algorithm may fail sometimes, however, but a slightly modified algorithm that converges quickly and monotonically is described by Vardi and Zhang (2001).

Next we consider the consistency and limiting distribution of the spatial median $\hat{\mu}(\mathbf{Y})$. Under general assumptions, if $\mathbf{Y}$ is a random sample from F, then the sample spatial median $\hat{\mu}(\mathbf{Y})$ converges to the population spatial median $\mu = \mu(F)$.

Definition 6.3. Assume that the cumulative distribution function of $\mathbf{y}_i$ is F. Then the theoretical or population spatial median $\mu = \mu(F)$ minimizes the criterion function

$$D(\mu) = \mathbf{E}\{|\mathbf{y}_i - \mu| - |\mathbf{y}_i|\}.$$

Note that, as $|\mathbf{y}_i - \mu| - |\mathbf{y}_i| \leq |\mu|$, the expectation in the above definition always exists.

For the asymptotical properties of the estimate we need the following assumption.

Assumption 1 *The density function of* $\mathbf{y}_i$ *is uniformly bounded and continuous. Moreover, the population spatial median* $\mu = \mu(F)$ *is unique; that is,* $D(\mu') > D(\mu)$ *for all* $\mu' \neq \mu$.

For the consideration of the limiting distribution, we can assume that the *population spatial median* is $\mathbf{0}$. This is not a restriction as the estimate and the functional are clearly *shift equivariant*. Note also that no assumption about the existence of the moments is needed.

First note that both $D_n(\mu)$ and $D(\mu)$ are convex and that, based on Lemmas 6.1 and 6.3, we have pointwise convergences

$$D_n(\mu) \to_P D(\mu) = \mu'\mathbf{A}\mu + o(|\mu|^2)$$

and

$$nD_n(n^{-1/2}\mu) - \left(\sqrt{n}\mathbf{T} - \frac{1}{2}\mathbf{A}\mu\right)'\mu = o_P(1).$$

Then Theorem B.5 in Appendix B gives the following result.

Theorem 6.5. *Under Assumption 1, $\hat{\mu} \to_P \mu$ and the limiting distribution of $\sqrt{n}(\hat{\mu} - \mu)$ is*

$$N_p\left(\mathbf{0}, \mathbf{A}^{-1}\mathbf{B}\mathbf{A}^{-1}\right).$$

The proofs can be found in Appendix B. Heuristically, the results in Theorem 6.5 can be simply based on Taylor's expansion with $\mu = \mathbf{0}$,

$$\mathbf{0} = \sqrt{n}\,\mathrm{\mathbf{AVE}}\{\mathbf{U}(\mathbf{y}_i - \hat{\mu})\} = \sqrt{n}\,\mathrm{\mathbf{AVE}}\{\mathbf{U}_i\} - \mathrm{\mathbf{AVE}}\{\mathbf{A}(\mathbf{y}_i)\}\sqrt{n}\hat{\mu} + o_P(1),$$

where

$$\mathbf{A}(\mathbf{y}) = |\mathbf{y}|^{-1}(\mathbf{I} - |\mathbf{y}|^{-2}\mathbf{y}\mathbf{y}')$$

is the $p \times p$ Hessian matrix (matrix of the second derivatives) of $|\mathbf{y}|$. The result follows as

$$\mathrm{\mathbf{AVE}}\{\mathbf{A}(\mathbf{y}_i)\} \to_P \mathbf{A}$$

and

$$\sqrt{n}\,\mathrm{\mathbf{AVE}}\{\mathbf{U}_i\} \to_d N_p(\mathbf{0}, \mathbf{B}).$$

For a practical use of the normal approximation of the distribution of $\hat{\mu}$ one naturally needs an estimate for the asymptotic covariance matrix $\mathbf{A}^{-1}\mathbf{B}\mathbf{A}^{-1}$. We estimate $\mathbf{A}$ and $\mathbf{B}$ separately. First, let

$$\mathbf{A}(\mathbf{y}) = |\mathbf{y}|^{-1}(\mathbf{I} - |\mathbf{y}|^{-2}\mathbf{y}\mathbf{y}') \quad \text{and} \quad \mathbf{B}(\mathbf{y}) = |\mathbf{y}|^{-2}\mathbf{y}\mathbf{y}'.$$

Then write

$$\hat{\mathbf{A}} = \mathrm{\mathbf{AVE}}\{\mathbf{A}(\mathbf{y}_i - \hat{\mu})\} \text{ and } \hat{\mathbf{B}} = \mathrm{\mathbf{AVE}}\{\mathbf{B}(\mathbf{y}_i - \hat{\mu})\},$$

which, under the stated assumption, converge in probability to the population values

$$\mathbf{A} = \mathbf{E}\{\mathbf{A}(\mathbf{y}_i - \mu)\} \quad \text{and} \quad \mathbf{B} = \mathbf{E}\{\mathbf{B}(\mathbf{y}_i - \mu)\},$$

respectively. We now prove the following

Theorem 6.6. *Under Assumption 1, $\hat{\mathbf{A}} \to_P \mathbf{A}$ and $\hat{\mathbf{B}} \to_P \mathbf{B}$.*

Proof. It is not a restriction to assume that $\mu = \mathbf{0}$. Then the spatial median $\hat{\mu}$ is a root-n consistent estimate of $\mathbf{0}$. We write

$$\tilde{\mathbf{A}} = \mathrm{\mathbf{AVE}}\{\mathbf{A}(\mathbf{y}_i)\} \quad \text{and} \quad \tilde{\mathbf{B}} = \mathrm{\mathbf{AVE}}\{\mathbf{B}(\mathbf{y}_i)\}.$$

Note that $\tilde{\mathbf{A}}$ and $\tilde{\mathbf{B}}$ would be our estimates for $\mathbf{A}$ and for $\mathbf{B}$, respectively, if we knew the true value $\mu = \mathbf{0}$. Then naturally

$$\tilde{\mathbf{A}} \to_P \mathbf{A} \quad \text{and} \quad \tilde{\mathbf{B}} \to_P \mathbf{B}.$$

As

$$\left| \frac{\mathbf{a}-\mathbf{b}}{|\mathbf{a}-\mathbf{b}|} - \frac{\mathbf{a}}{|\mathbf{a}|} \right| \le 2\frac{|\mathbf{b}|}{|\mathbf{a}|}, \quad \forall\, \mathbf{a} \neq 0, \mathbf{b}$$

and

$$\left| \frac{(\mathbf{a}-\mathbf{b})(\mathbf{a}-\mathbf{b})}{|\mathbf{a}-\mathbf{b}|^2} - \frac{\mathbf{a}\mathbf{a}'}{|\mathbf{a}|^2} \right| \le 4\frac{|\mathbf{b}|}{|\mathbf{a}|}, \quad \forall\, \mathbf{a} \neq 0, \mathbf{b}$$

then

$$|\hat{\mathbf{B}} - \tilde{\mathbf{B}}| \le \frac{1}{n}\sum_{i=1}^{n} 4\frac{|\hat{\mu}|}{|\mathbf{y}_i|} \to_P 0.$$

(Use Slutsky's theorem.) As $\tilde{\mathbf{B}} \to_P \mathbf{B}$, also $\hat{\mathbf{B}} \to_P \mathbf{B}$.

It is much trickier to prove that $\hat{\mathbf{A}} \to_P \mathbf{A}$. We need to play with three positive constants δ_1, δ_2, and δ_3. We use the root-n consistency of $\hat{\mu}$ and assume that $|\hat{\mu}| < \delta_1/\sqrt{n}$. (This is true with a probability that can be made close to one with large δ_1.) Also the observations $\mathbf{y}_i$ should somehow be blocked out from $\hat{\mu}$, $i = 1,...,n$. For that, write

$$I_{1i} = I\left\{ |\mathbf{y}_i - \hat{\mu}| < \frac{\delta_2}{\sqrt{n}} \right\},$$

$$I_{2i} = I\left\{ \frac{\delta_2}{\sqrt{n}} \le |\mathbf{y}_i - \hat{\mu}| < \delta_3 \right\}, \text{ and}$$

$$I_{3i} = I\{ |\mathbf{y}_i - \hat{\mu}| \ge \delta_3 \}.$$

Then

$$\begin{aligned}
\tilde{\mathbf{A}} - \hat{\mathbf{A}} = {} & \frac{1}{n}\sum_{i=1}^{n} (\mathbf{A}(\mathbf{y}_i) - \mathbf{A}(\mathbf{y}_i - \hat{\mu})) \\
= {} & \frac{1}{n}\sum_{i=1}^{n} (I_{1i} \cdot [\mathbf{A}(\mathbf{y}_i) - \mathbf{A}(\mathbf{y}_i - \hat{\mu})]) \\
& + \frac{1}{n}\sum_{i=1}^{n} (I_{2i} \cdot [\mathbf{A}(\mathbf{y}_i) - \mathbf{A}(\mathbf{y}_i - \hat{\mu})]) \\
& + \frac{1}{n}\sum_{i=1}^{n} (I_{3i} \cdot [\mathbf{A}(\mathbf{y}_i) - \mathbf{A}(\mathbf{y}_i - \hat{\mu})]).
\end{aligned}$$

The first average, with nonzero terms in a shrinking neighborhood of $\hat{\mu}$ only, is zero with a probability

$$P(I_{11} = \cdots = I_{1n} = 0) \ge \left(1 - \frac{\delta_2^p cM}{n^{p/2}}\right)^n \ge \left(1 - \frac{\delta_2^2 cM}{n}\right)^n \to e^{-cM\delta_2^2},$$

where $M = \sup_{\mathbf{y}} f(\mathbf{y}) < \infty$ and $c = \pi^{p/2}/\Gamma((p+3)/2)$ is the volume of the p-variate unit ball. The first average is thus zero with a probability that can be made close to one with small choices of $\delta_2 > 0$. For the second average, one gets

$$\frac{1}{n}\sum_{i=1}^{n} |I_{2i} \cdot [\mathbf{A}(\mathbf{y}_i) - \mathbf{A}(\mathbf{y}_i - \hat{\mu})]| \leq \frac{1}{n}\sum_{i=1}^{n} \frac{6 I_{2i}|\hat{\mu}|}{|\mathbf{y}_i - \hat{\mu}||\mathbf{y}_i|}$$
$$\leq \frac{1}{n}\sum_{i=1}^{n} \frac{6 I_{2i}\delta_1}{\delta_2|\mathbf{y}_i|}$$

which converges to a constant that can be made as close to zero as one wishes with small $\delta_3 > 0$. Finally also the third average

$$\frac{1}{n}\sum_{i=1}^{n} |I_{3i} \cdot [\mathbf{A}(\mathbf{y}_i) - \mathbf{A}(\mathbf{y}_i - \hat{\mu})]| \leq \frac{1}{n}\sum_{i=1}^{n} \frac{6 I_{3i}|\hat{\mu}|}{|\mathbf{y}_i - \hat{\mu}||\mathbf{y}_i|}$$
$$\leq \frac{1}{n\sqrt{n}}\sum_{i=1}^{n} \frac{6 I_{3i}\delta_1}{\delta_3|\mathbf{y}_i|}$$

converges to zero in probability for all choices of δ_1 and δ_3.

Theorems 7.4 and 6.6 thus imply that the distribution of $\hat{\mu}$ can be approximated by

$$N_p\left(\mu, \frac{1}{n}\hat{\mathbf{A}}^{-1}\hat{\mathbf{B}}\hat{\mathbf{A}}^{-1}\right),$$

and approximate confidence ellipsoids for μ can be constructed. In the spherically symmetric case the estimation is much easier as the matrices are simply $(p > 1)$

$$\mathbf{A} = \frac{(p-1)\mathbf{E}[|\mathbf{y}_i - \mu|^{-1}]}{p}\,\mathbf{I}_p \quad \text{and} \quad \mathbf{B} = \frac{1}{p}\,\mathbf{I}_p.$$

An estimate of the limiting covariance matrix of the spatial median is then

$$\frac{p}{(p-1)^2}\frac{1}{[\mathbf{AVE}\{|\mathbf{y}_i - \hat{\mu}|^{-1}\}]^2}\mathbf{I}_p.$$

The spatial median is extremely robust: Brown showed that the estimator has a bounded influence function. It also has a breakdown point of $1/2$. See Niinimaa and Oja (1995) and Lopuhaä and Rousseeuw (1991).

If all components are on the same unit of measurement (and all the components may be rescaled only in a similar way), the spatial median is an attractive descriptive measure of location. Rotating the data cloud rotates the median correspondingly; that is,

$$\hat{\mu}(\mathbf{YO}') = \mathbf{O}\hat{\mu}(\mathbf{Y}).$$

Unfortunately, the estimate is not equivariant under arbitrary affine transformations.

6.2.2 The estimate with inner standardization

To create an affine equivariant version of the multivariate median, we transform the data as we did in the test construction. The estimates, however, must be transformed back to the original coordinate system. The procedure is then as follows.

1. Take any scatter matrix $\mathbf{S} = \mathbf{S}(\mathbf{Y})$.
2. Standardize the data matrix: $\mathbf{YS}^{-1/2}$.
3. Find the spatial median for the standardized data matrix $\hat{\mu}(\mathbf{YS}^{-1/2})$.
4. Retransform the estimate: $\tilde{\mu}(\mathbf{Y}) = \mathbf{S}^{1/2}\hat{\mu}(\mathbf{YS}^{-1/2})$.

This median utilizing "data-driven" transformation $\mathbf{S}^{-1/2}$ is known as the *transformation-retransformation (TR) spatial median*, and was considered by Chakraborty et al. (1998). Then the affine equivariance follows.

Theorem 6.7. *Let* $\mathbf{S} = \mathbf{S}(\mathbf{Y})$ *be any scatter matrix. Then the transformation retransformation spatial median*

$$\tilde{\mu}(\mathbf{Y}) = \mathbf{S}^{1/2}\hat{\mu}(\mathbf{YS}^{-1/2})$$

is affine equivariant.

It is remarkable that the almost sure convergence and the limiting normality of the spatial median did not require any moment assumptions. Therefore, for the transformation, a scatter matrix with weak assumptions should be used as well. It is an appealing idea also to link the spatial median with Tyler's transformation. This was proposed Hettmansperger and Randles (2002).

Definition 6.4. Let μ be a p-vector and $\mathbf{S} > 0$ a symmetric $p \times p$ matrix, and define

$$\varepsilon_i = \varepsilon_i(\mu, \mathbf{S}) = \mathbf{S}^{-1/2}(\mathbf{y}_i - \mu), \quad i = 1, ..., n.$$

The Hettmansperger-Randles (HR) estimate of location and scatter are the values of μ and $\mathbf{S}$ that simultaneously satisfy

$$\mathbf{AVE}\{\mathbf{U}(\varepsilon_i)\} = \mathbf{0} \quad \text{and} \quad p\,\mathbf{AVE}\{\mathbf{U}(\varepsilon_i)\mathbf{U}(\varepsilon_i)'\} = \mathbf{I}_p.$$

In the HR estimation, the location estimate is the TR spatial median, and the scatter estimate is Tyler's estimate with respect to the TR spatial median. The shift vector and scatter matrix are thus obtained using inner centering and standardization with the spatial sign score function $\mathbf{U}(\mathbf{y})$. The location and scatter estimates are affine equivariant and apparently estimate μ and Σ in the model (B2) of elliptical directions. Their properties were developed by Hettmansperger and Randles (2002). They showed that the HR estimate has a bounded influence function and a positive breakdown point. The limiting distribution is given by the following.

Theorem 6.8. *Let* $\mathbf{Y} = (\mathbf{y}_1, ..., \mathbf{y}_n)'$ *be a random sample and assume that the* $\mathbf{y}_i$ *are generated by*

$$\mathbf{y}_i = \Omega \varepsilon_i + \mu, \quad i = 1, ..., n,$$

where

$$\mathbf{E}\{\mathbf{U}(\varepsilon_i)\} = \mathbf{0} \quad \text{and} \quad p\,\mathbf{E}\{\mathbf{U}(\varepsilon_i)\mathbf{U}(\varepsilon_i)'\} = \mathbf{I}_p.$$

Then the limiting distribution of $\sqrt{n}(\tilde{\mu} - \mu)$ *is* $N_p(\mathbf{0}, p^{-1}\mathbf{S}^{1/2}\mathbf{A}^{-2}\mathbf{S}^{1/2})$ *where* $\mathbf{A} = E(\mathbf{A}(\mathbf{S}^{-1/2}\mathbf{y}_i))$ *and* $\mathbf{S}$ *is Tyler's scatter matrix.*

The HR estimate is easy to compute even in high dimensions. The iteration steps (as in M-estimation) first update the residuals, then the location center, and finally the scatter matrix as follows.

1.
$$\varepsilon_i \leftarrow \mathbf{S}^{-1/2}(\mathbf{y}_i - \mu), \quad i = 1, ..., n.$$

2.
$$\mu \leftarrow \mu + \frac{\mathbf{S}^{1/2}\,\mathbf{AVE}\{\mathbf{U}(\varepsilon_i)\}}{\mathbf{AVE}\{|\varepsilon_i|^{-1}\}}.$$

3.
$$\mathbf{S} \leftarrow p\,\mathbf{S}^{1/2}\,\mathbf{AVE}\{\mathbf{U}(\varepsilon_i)\mathbf{U}(\varepsilon_i)'\}\,\mathbf{S}^{1/2}.$$

Unfortunately, there is no proof so far for the convergence of the above algorithm although in practice it always seems to work. There is no proof for the existence or uniqueness of the HR estimate either. In practice, this is not a problem, however. If, in the spherical case around the origin, the initial location and shape estimates, say $\mathbf{M}$ and $\mathbf{S}$ are root-n consistent, that is,

$$\sqrt{n}\mathbf{M} = O_P(1) \quad \text{and} \quad \sqrt{n}(\mathbf{S} - \mathbf{I}_p) = O_P(1)$$

and $tr(\mathbf{S}) = p$, then the k-step estimates (obtained after k iterations of the above algorithm) satisfy

$$\sqrt{n}\mathbf{M}_k = \left(\frac{1}{p}\right)^k \sqrt{n}\mathbf{M} + \left[1 - \left(\frac{1}{p}\right)^k\right] \frac{1}{E(r_i^{-1})} \frac{p}{p-1} \sqrt{n}\mathbf{AVE}\{\mathbf{u}_i\} + o_P(1)$$

and

$$\sqrt{n}(\mathbf{S}_k - \mathbf{I}_p) = \left(\frac{2}{p+2}\right)^k \sqrt{n}(\mathbf{S} - \mathbf{I}_p)$$
$$+ \left[1 - \left(\frac{2}{p+2}\right)^k\right] \frac{p+2}{p} \sqrt{n}\left(p \cdot \mathbf{AVE}\{\mathbf{u}_i\mathbf{u}_i'\} - \mathbf{I}_p\right) + o_P(1).$$

Asymptotically, the k-step estimate behaves as a linear combination of the initial pair of estimates and Hettmansperger-Randles estimate. The larger k is, the more similar is the distribution to that of the HR estimate.

Example 6.5. **Cork boring data.** If we then wish to estimate the unknown spatial median (3-variate case), then the regular spatial median and HR estimate behave in a quite similar way. See also Figure 6.5.

```
> est.sign.o.3v <- mv.1sample.est(cork_3v, "s")
> summary(est.sign.o.3v)
The spatial median of cork_3v is:
[1] -3.5013 -0.0875 -3.9750

And has the covariance matrix:
        [,1]    [,2]    [,3]
[1,] 2.0120 0.9025 0.4734
[2,] 0.9025 3.3516 1.3180
[3,] 0.4734 1.3180 3.3704
>
> est.sign.i.3v <- mv.1sample.est(cork_3v, "s", "i")
> summary(est.sign.i.3v)
The equivariant spatial median of cork_3v is:
[1] -3.2736 -0.0013 -4.2687

And has the covariance matrix:
       [,1]  [,2]  [,3]
[1,] 2.172 1.374 0.154
[2,] 1.374 3.487 1.297
[3,] 0.154 1.297 2.364
>
> plotMvloc(est.sign.o.3v, est.sign.i.3v, X=cork_3v,
  color.ell=1:3, , lty.ell=1:3, pch.ell= 15:17)
```

Example 6.6. **Cork boring data.** In the bivariate case we get

```
> summary(est.sign.o.2v)
The spatial median of cork_2v is:
[1] -0.3019  0.0580

And has the covariance matrix:
        [,1]    [,2]
```

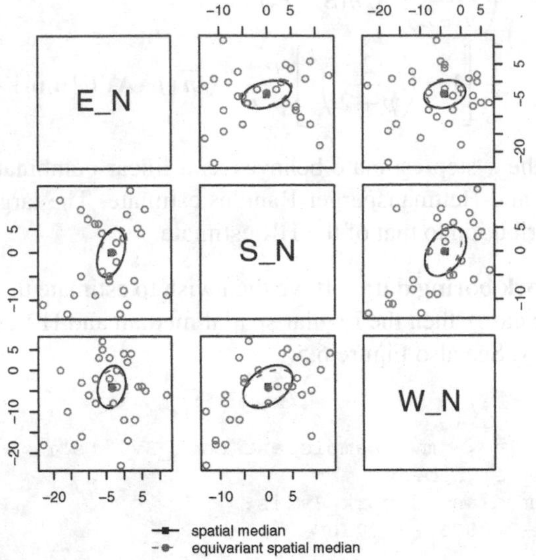

Fig. 6.5 The scatterplot with estimated regular spatial median and the HR estimate and their 95% confidence ellipsoids.

```
[1,]  5.434 -1.883
[2,] -1.883  6.184
>
> est.sign.i.2v <- mv.1sample.est(cork_2v, "s", "i")
> summary(est.sign.i.2v)
The equivariant spatial median of cork_2v is:
[1] -0.2467 -0.0234

And has the covariance matrix:
       [,1]   [,2]
[1,]  5.626 -1.595
[2,] -1.595  6.017
>
> plotMvloc(est.sign.o.2v, est.sign.i.2v, X=cork_2v,
 color.ell=1:3, , lty.ell=1:3, pch.ell= 15:17)
```

The regular spatial median and HR estimate behave almost identically in the case of bivariate data; see Figure 6.6. However, if the second component is multiplied by 10 (Figure 6.7), the results differ. The equivariant spatial median now has a smaller confidence ellipsoid because the regular spatial median loses efficiency if the marginal variables are heterogeneous in their variation. The R code for this comparison is as follows.

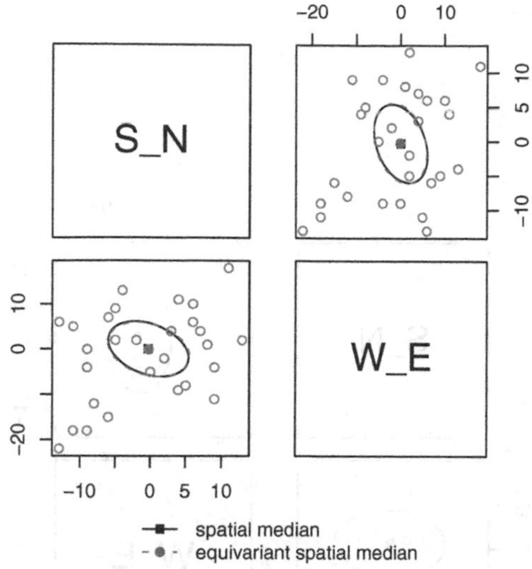

Fig. 6.6 The scatterplot with estimated regular spatial median and the HR estimate and their 95% confidence ellipsoids.

```
> cork_2v_scaled <- transform(cork_2v, W_E = W_E * 10)
>
> est.sign.o.2v_scaled <- mv.1sample.est(cork_2v_scaled, "s")
> summary(est.sign.o.2v_scaled)
The spatial median of cork_2v_scaled is:
[1]  0.4728 12.7727

And has the covariance matrix:
       [,1]    [,2]
[1,] 11.161   4.616
[2,]  4.616 880.176
>
> est.sign.i.2v_scaled <-
 mv.1sample.est(cork_2v_scaled, "s", "i")
> summary(est.sign.i.2v_scaled)
The equivariant spatial median of cork_2v_scaled is:
[1] -0.2467 -0.2337

And has the covariance matrix:
        [,1]    [,2]
[1,]   5.626 -15.94
[2,] -15.945 601.74
>
> plotMvloc(est.sign.o.2v_scaled, est.sign.i.2v_scaled,
```

```
X=cork_2v_scaled, color.ell=1:3, , lty.ell=1:3,
pch.ell= 15:17)
```

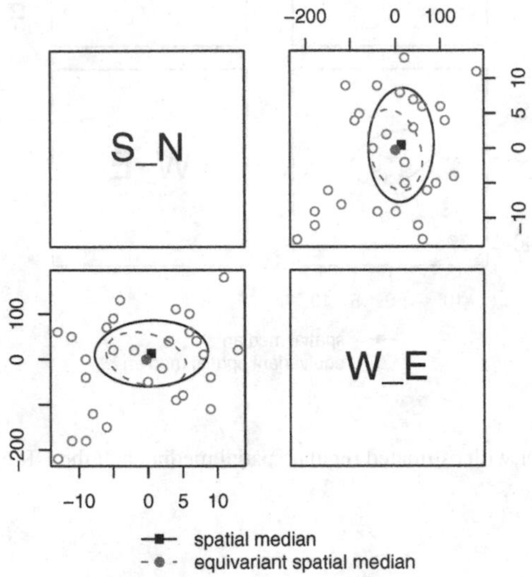

Fig. 6.7 The estimates with 95 % confidence ellipsoids for the regular spatial median and the affine equivariant spatial median (HR estimate).

6.2.3 Other multivariate medians

The *vector of marginal medians* minimizes the L_1 objective function

$$\mathbf{AVE}\{|y_{i1}-\mu_1|+\cdots+|y_{ip}-\mu_p|\}.$$

See Puri and Sen (1971) and Rao (1988) for the asymptotic covariance matrix of the vector of marginal sample medians. The asymptotic efficiencies naturally agree with the univariate asymptotic efficiencies. It is not affine invariant but the transformation-retransformation technique can be used to find the invariant version of the estimate; see Chakraborty and Chaudhuri (1998).

There are a number of affine equivariant multivariate generalizations of the median: the *half-space median* Tukey (1975), the *multivariate Oja median* Oja (1983), and the *multivariate Liu median* Liu (1990). For these and other multivariate medians, see the surveys by Small (1990) and Niinimaa and Oja (1999).

Chapter 7
One-sample problem: Spatial signed-rank test and Hodges-Lehmann estimate

Abstract The spatial signed-rank score function $\mathbf{Q}(\mathbf{y})$ is used for the one-sample location problem. The test is then the spatial signed-rank test, and the estimate is the spatial Hodges-Lehmann estimate. The tests and estimates based on outer standardization as well as those based on inner standardization are again discussed.

7.1 Multivariate spatial signed-rank test

We consider first the location model

$$\mathbf{y}_i = \boldsymbol{\mu} + \boldsymbol{\varepsilon}_i, \quad i = 1, ..., n,$$

where the independent residuals $\boldsymbol{\varepsilon}_i$ have a joint density $f(\boldsymbol{\varepsilon})$ that is uniformly bounded. The residuals are now thought to be centered so that

$$\mathbf{E}(\mathbf{U}(\boldsymbol{\varepsilon}_i + \boldsymbol{\varepsilon}_j)) = \mathbf{0}, \quad i \neq j.$$

As before, the cumulative distribution function of $\boldsymbol{\varepsilon}_i$ is denoted by $F(\boldsymbol{\varepsilon})$. In this model, parameter $\boldsymbol{\mu}$ is the so-called Hodges-Lehmann center of the distribution of $\mathbf{y}_i$; that is,

$$\mathbf{E}(\mathbf{U}(\mathbf{y}_i + \mathbf{y}_j - 2\boldsymbol{\mu})) = \mathbf{0}, \quad i \neq j.$$

Parameter $\boldsymbol{\mu}$ is the spatial median of the distribution of the average $(\mathbf{y}_1 + \mathbf{y}_2)/2$. If F is symmetrical then naturally $\boldsymbol{\mu}$ is the symmetry center. We wish to test the null hypothesis

$$H_0: \ \boldsymbol{\mu} = \mathbf{0}.$$

Again, we start by giving the (theoretical) score function $\mathbf{Q}_F$, the test statistic $\mathbf{T}(\mathbf{Y})$, and the matrices $\mathbf{A}$ and $\mathbf{B}$. The multivariate spatial signed-rank test is obtained if one uses the spatial signed-rank score function

H. Oja, *Multivariate Nonparametric Methods with R: An Approach Based on Spatial Signs and Ranks*, Lecture Notes in Statistics 199, DOI 10.1007/978-1-4419-0468-3_7,

$$\mathbf{T}(\mathbf{y}) = \mathbf{Q}_F(\mathbf{y}) = \frac{1}{2}\mathbf{E}\{\mathbf{U}(\mathbf{y}-\varepsilon_i)+\mathbf{U}(\mathbf{y}+\varepsilon_i)\}.$$

Then the test for testing $H_0: \mu = \mathbf{0}$ uses

$$\mathbf{T}(\mathbf{Y}) = \mathbf{AVE}\{\mathbf{Q}_F(\mathbf{y}_i)\}.$$

The asymptotical properties of the test are based on the matrices

$$\mathbf{A} = \mathbf{E}\{\mathbf{A}(\varepsilon_1+\varepsilon_2)\} \quad \text{and} \quad \mathbf{B} = \mathbf{QCOV}(F) = \mathbf{E}(\mathbf{Q}_F(\varepsilon_i)\mathbf{Q}_F(\varepsilon_i)'),$$

where as before $\mathbf{A}(\varepsilon) = |\varepsilon|^{-1}(\mathbf{I}_p - |\varepsilon|^{-2}\varepsilon\varepsilon')$. As the density function of ε_i is assumed to be continuous and uniformly bounded $\mathbf{A}$ also exists.

Note that the population signed-rank score function is of course unknown in practice and it is in the test construction replaced by the estimated score function

$$\mathbf{Q}(\mathbf{y}) = \frac{1}{2}\mathbf{AVE}\{\mathbf{U}(\mathbf{y}-\mathbf{y}_i)+\mathbf{U}(\mathbf{y}+\mathbf{y}_i)\}.$$

($\mathbf{Q}(\mathbf{y})$ is a consistent estimate of $\mathbf{Q}_F(\mathbf{y})$ if the null hypothesis is true.) Then the observed *spatial signed-ranks* are then

$$\mathbf{Q}_i = \mathbf{Q}(\mathbf{y}_i), \quad i = 1,...,n.$$

and the test statistic using the estimated scores is simply the following.

Definition 7.1. The *spatial signed-rank test statistic* for testing $H_0: \mu = \mathbf{0}$ is the average of spatial signed-ranks,

$$\hat{\mathbf{T}}(\mathbf{Y}) = \mathbf{AVE}\{\mathbf{Q}_i\}.$$

As $\mathbf{AVE}\{\mathbf{U}(\mathbf{y}_i-\mathbf{y}_j)\} = \mathbf{0}$ the test statistic can be simplified to

$$\hat{\mathbf{T}}(\mathbf{Y}) = \frac{1}{2}\mathbf{AVE}\{\mathbf{U}(\mathbf{y}_i+\mathbf{y}_j)\}.$$

The statistic $\hat{\mathbf{T}}$ is a V-statistic and asymptotically equivalent to the corresponding U-statistic

$$\tilde{\mathbf{T}}(\mathbf{Y}) = \frac{1}{2}\mathbf{AVE}_{i<j}\{\mathbf{U}(\mathbf{y}_i-\mathbf{y}_j)+\mathbf{U}(\mathbf{y}_i+\mathbf{y}_j)\}.$$

For the theory of U-statistics and V-statistics, we refer to Serfling (1980). Asymptotic equivalence means that

$$\sqrt{n}\left(\tilde{\mathbf{T}}(\mathbf{Y}) - \hat{\mathbf{T}}(\mathbf{Y})\right) \to_P \mathbf{0}.$$

The test statistic $\hat{\mathbf{T}}$ is only orthogonal equivariant, not affine equivariant. Moreover, its finite sample and asymptotic null distribution depend both on the distributions of modulus $|\mathbf{y}_i|$ and on the distribution of the direction $\mathbf{U}_i$. A natural estimate of its asymptotic covariance matrix is the signed-rank covariance matrix,

$$\hat{\mathbf{B}} = \mathbf{B}(\mathbf{Y}) = \mathbf{QCOV}(\mathbf{Y}) = \mathbf{AVE}\left\{\mathbf{Q}_i\mathbf{Q}_i'\right\}.$$

Note that $\hat{\mathbf{B}}$ is asymptotically equivalent with a U-statistic with symmetric and bounded kernel and the next lemma easily follows.

Lemma 7.1. *Under H_0, $\hat{\mathbf{B}} \rightarrow_P \mathbf{B}$.*

We thus replace the "true" but unknown test statistic $\mathbf{T}(\mathbf{Y})$ by test statistic $\hat{\mathbf{T}}(\mathbf{Y})$ which in turn is asymptotically equivalent with $\tilde{\mathbf{T}}(\mathbf{Y})$. It is straightforward to see that $\mathbf{T}(\mathbf{Y})$ is the projection of $\tilde{\mathbf{T}}(\mathbf{Y})$ in the sense that

$$\mathbf{T}(\mathbf{Y}) = \sum_{i=1}^{n} E[\tilde{\mathbf{T}}(\mathbf{Y})|\mathbf{y}_i]$$

and therefore (use Theorem 5.3.2 in Serfling (1980) with a bounded kernel) the following holds.

Lemma 7.2. $\sqrt{n}[\mathbf{T}(\mathbf{Y}) - \tilde{\mathbf{T}}(\mathbf{Y})] \rightarrow_P 0$ *and also* $\sqrt{n}[\mathbf{T}(\mathbf{Y}) - \hat{\mathbf{T}}(\mathbf{Y})] \rightarrow_P 0$.

The regular central limit theorem (CLT) with independent and identically distributed observations then gives for $\hat{\mathbf{T}} = \hat{\mathbf{T}}(\mathbf{Y})$

Theorem 7.1. *Under $H_0:\ \mu = \mathbf{0}$, $\sqrt{n}\hat{\mathbf{T}} \rightarrow_d N(\mathbf{0},\mathbf{B})$ and*

$$Q^2 = Q^2(\mathbf{Y}) = n\hat{\mathbf{T}}'\hat{\mathbf{B}}^{-1}\hat{\mathbf{T}} \rightarrow \chi_p^2.$$

Totally nonparametric model. We collect the results obtained so far. We first transform

$$\mathbf{Y} = (\mathbf{y}_1,...,\mathbf{y}_n)' \quad \rightarrow \quad \mathbf{Q} = (\mathbf{Q}_1,...,\mathbf{Q}_n)'.$$

We then use outer standardization and get (the squared form of) the test statistic

$$Q^2 = \mathbf{1}_n'\mathbf{Q}(\mathbf{Q}'\mathbf{Q})^{-1}\mathbf{Q}'\mathbf{1}_n \quad \text{or} \quad n \cdot tr\left(\mathbf{Q}'\mathbf{P}_{\mathbf{1}_n}\mathbf{Q}(\mathbf{Q}'\mathbf{Q})^{-1}\right).$$

If the distribution of $\mathbf{y}_i$ is centered around μ in the sense that

$$\mathbf{E}(\mathbf{U}(\mathbf{y}_1 + \mathbf{y}_2 - 2\mu)) = \mathbf{0},$$

then the limiting distribution of the test statistic Q^2 under $H_0:\ \mu = \mathbf{0}$ is a chi square distribution with p degrees of freedom and the approximate p-values are found as the tail probabilities of χ_p^2.

Recall that Hotelling's test and the test based on spatial signs were similarly

$$\mathbf{1}_n'\mathbf{Y}(\mathbf{Y}'\mathbf{Y})^{-1}\mathbf{Y}'\mathbf{1}_n \quad \text{and} \quad \mathbf{1}_n'\mathbf{U}(\mathbf{U}'\mathbf{U})^{-1}\mathbf{U}'\mathbf{1}_n,$$

respectively.

Nonparametric model with symmetry assumptions. An exact sign-change test version is obtained if

$$\mathbf{JY} \sim \mathbf{Y}$$

for all $n \times n$ sign-change matrices $\mathbf{J}$. Note that $\mathbf{QCOV}(\mathbf{JY}) = \mathbf{QCOV}(\mathbf{Y})$ and

$$Q^2(\mathbf{JY}) = \mathbf{1}_n'\mathbf{JQ}(\mathbf{Q}'\mathbf{Q})^{-1}\mathbf{Q}'\mathbf{J}\mathbf{1}_n$$

and the exact p-value for the conditionally distribution-free sign-change test is

$$\mathbf{E_J}\left[I\left\{Q^2(\mathbf{JY}) \geq Q^2(\mathbf{Y})\right\}\right].$$

Consider next the model

$$\mathbf{y}_i = \Omega\varepsilon_i + \mu, \quad i = 1,...,n,$$

where the standardized residuals ε_i are independent, symmetric, and centered so that

$$\mathbf{E}(\mathbf{U}(\varepsilon_i + \varepsilon_j)) = \mathbf{0}, \quad i \neq j.$$

Symmetry center μ is thus the Hodges-Lehmann center of the distribution of $\mathbf{y}_i$. As the model is closed under affine transformations, it is natural to require that the test is affine invariant (the selected coordinate system does not have an effect on the p-value obtained from the test). The signed-rank test statistic is not affine invariant. However, we again have the following result.

Theorem 7.2. *Let $\mathbf{S} = \mathbf{S}(\mathbf{Y})$ be any scatter matrix. Then the signed-rank test statistic calculated for the transformed data set, $Q^2(\mathbf{YS}^{-1/2})$, is affine invariant.*

A natural choice for $\mathbf{S}$ is the scatter matrix that makes the signed-rank covariance matrix of the standardized observations proportional to the identity matrix. Let us give the following definition.

Definition 7.2. The *scatter matrix based on signed-ranks* is the symmetric $p \times p$ matrix $\mathbf{S} = \mathbf{S}(\mathbf{Y}) > 0$ with $tr(\mathbf{S}) = p$ such that, if $\hat{\mathbf{Q}}_i = \mathbf{Q}_{\mathbf{YS}^{-1/2}}(\mathbf{S}^{-1/2}\mathbf{y}_i)$ then the signed-ranks $\hat{\mathbf{Q}}_i$ satisfy

$$p \cdot \mathbf{AVE}\left\{\hat{\mathbf{Q}}_i\hat{\mathbf{Q}}_i'\right\} = \mathbf{AVE}\left\{|\hat{\mathbf{Q}}_i|^2\right\} \cdot \mathbf{I}_p.$$

The transformation $\mathbf{S}^{-1/2}$ makes the signed-rank covariance matrix $\mathbf{QCOV}(\mathbf{YS}^{-1/2})$ proportional to the identity matrix as if the signed-ranks were spherically distributed on the unit p-ball $\mathscr{B}_p$. The iterative algorithm for its computation again uses the iteration steps

$$\mathbf{S} \leftarrow \mathbf{S}^{1/2}\mathbf{QCOV}(\mathbf{YS}^{-1/2})\,\mathbf{S}^{1/2} \quad \text{and} \quad \mathbf{S} \leftarrow \frac{p}{tr(\mathbf{S})}\mathbf{S}.$$

Unfortunately, unlike for Tyler's scatter matrix, there is no proof of the convergence of the algorithm so far, but in practice it seems always to converge.

The spatial signed-ranks of the transformed observations, $\hat{\mathbf{Q}}_i$, $i = 1,...,n$, are called *standardized spatial signed-ranks*. The multivariate *spatial signed-rank test based on the inner standardization* then rejects H_0 for large values of

$$Q^2(\mathbf{Y}\mathbf{S}^{-1/2}) = \mathbf{1}_n'\hat{\mathbf{Q}}(\hat{\mathbf{Q}}'\hat{\mathbf{Q}})^{-1}\hat{\mathbf{Q}}'\mathbf{1}_n = np \cdot \frac{|\mathbf{AVE}\{\hat{\mathbf{Q}}_i\}|^2}{\mathbf{AVE}\{|\hat{\mathbf{Q}}_i|^2\}},$$

which is simply np times the ratio of the squared length of the average signed-rank to the average of squared lengths of signed-ranks. As in the case of the spatial sign test, we have the next theorem.

Theorem 7.3. *Test statistic $Q^2(\mathbf{Y}\mathbf{S}^{-1/2})$ is affine invariant and, under the null hypothesis $H_0: \mu = \mathbf{0}$,*

$$Q^2(\mathbf{Y}\mathbf{S}^{-1/2}) \to_d \chi^2_p.$$

Elliptically symmetric model. Consider the elliptical model

$$\mathbf{y}_i = \Omega\, \varepsilon_i + \mu, \quad i = 1,...,n,$$

where ε_i has a spherical distribution with cumulative distribution function F. Write $\varepsilon = (\varepsilon_1,...,\varepsilon_n)'$. Now the statistics $Q^2(\mathbf{Y}\mathbf{S}^{-1/2})$ and $Q^2(\varepsilon)$ are asymptotically equivalent. Let us construct the test using observations $(\varepsilon_1,...,\varepsilon_n)$ from a spherically symmetric distribution (model (A2)). Then

$$\mathbf{B} = \mathbf{QCOV}(F) = \mathbf{RCOV}(F) = \frac{E(q_F^2(|\varepsilon_i|))}{p}\mathbf{I}_p = \frac{\tau^2}{p}\mathbf{I}_p,$$

where constant τ^2 depends on the distribution of $|\varepsilon_i|$. But then the test statistic $Q^2(\mathbf{Y}\mathbf{S}^{-1/2})$ is asymptotically equivalent to

$$\frac{p}{4n^3\tau^2}\sum \cos(\varepsilon_i + \varepsilon_j, \varepsilon_{i'} + \varepsilon_{j'}),$$

where i, j, i', and j' all go over indices $1,...,n$. Jan and Randles (1994) constructed an affine invariant analogue of this test based again on the interdirection counts.

Example 7.1. **Cork boring data** Consider again the datasets with a 3-variate vector of E-N, S-N and W-N. The standardized spatial signed-ranks are illustrated in Figures 7.1. The observed value of $Q^2(\mathbf{Y})$, with inner standardization, in the 3-variate case is 13.67 and the corresponding p-value is 0.003.

```
> signed_ranks_i_3v <- spatial.signrank(cork_3v, FALSE, TRUE)
>
> pairs(signed_ranks_i_3v, labels = colnames(cork_3v), las = 1)
>
> mv.1sample.test(cork_3v, score = "r", stand = "i")

        One sample spatial signed-rank test using
        inner standardization

data:  cork_3v
Q.2 = 13.67, df = 3, p-value = 0.003384
alternative hypothesis: true location is not equal to c(0,0,0)
```

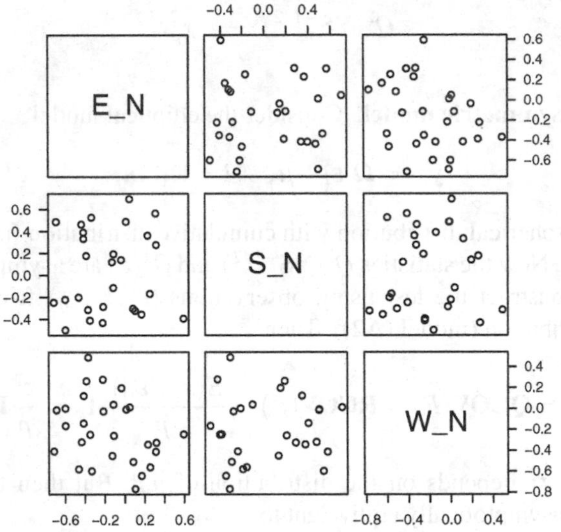

Fig. 7.1 The standardized spatial signed-ranks for the 3-variate data.

Next consider the bivariate data with variables S-N and W-E. The standardized spatial signed-ranks are illustrated in Figure 7.2. The observed value of $Q^2(\mathbf{Y})$ with inner standardization is 0.44 with corresponding p-value 0.80.

```
> signed_ranks_i_2v <- spatial.signrank(cork_2v, FALSE, TRUE)
>
```

```
> plot(signed_ranks_i_2v, xlab = "S_N", ylab = "W_E",
  ylim = c(-1, 1), xlim = c(-1, 1), las = 1, pty = "s")
>
> mv.1sample.test(cork_2v, score = "r", stand = "i")

        One sample spatial signed-rank test using
        inner standardization

data:  cork_2v
Q.2 = 0.4373, df = 2, p-value = 0.8036
alternative hypothesis: true location is not equal to c(0,0)
```

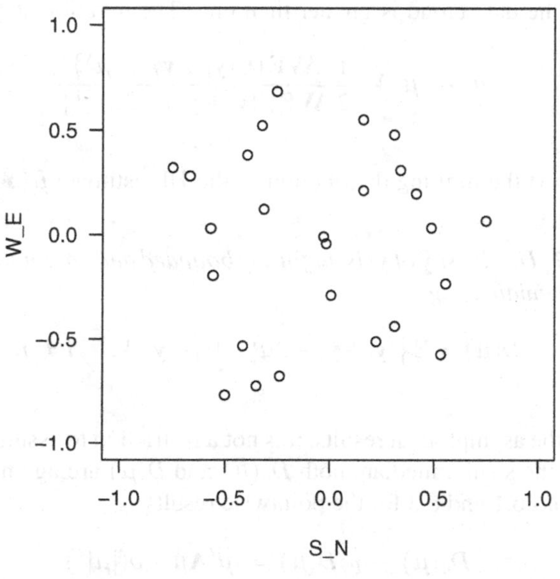

Fig. 7.2 The standardized spatial signed-ranks for the 2-variate data.

7.2 Multivariate spatial Hodges-Lehmann estimate

We now move to the estimation problem and define the *multivariate Hodges-Lehmann estimate* of location center μ as the spatial median of all pairwise means, Walsh averages

$$\frac{\mathbf{y}_i+\mathbf{y}_j}{2}, \quad i,j=1,...,n.$$

We thus define the following.

Definition 7.3. The *sample spatial Hodges-Lehmann (HL) estimate* $\hat{\mu}(\mathbf{Y})$ minimizes the criterion function

$$D_n(\mu) = \mathbf{AVE}\left\{|\mathbf{y}_i + \mathbf{y}_j - 2\mu| - |\mathbf{y}_i + \mathbf{y}_j|\right\}.$$

The link between the HL estimate and the signed-rank test statistic is again that $\hat{\mu}$ often solves the equation

$$\mathbf{AVE}\{\mathbf{U}(\mathbf{y}_i + \mathbf{y}_j - 2\hat{\mu})\} = \mathbf{0}.$$

The estimate $\hat{\mu}$ as a null value is the value with the highest possible p-value produced by the spatial signed-rank test. Again, the spatial median is unique, if the dimension of the data cloud is greater than one. The iteration step to compute its value is

$$\mu \leftarrow \mu + \frac{1}{2}\frac{\mathbf{AVE}\{\mathbf{U}(\mathbf{y}_i + \mathbf{y}_j - 2\mu)\}}{\mathbf{AVE}\{|(\mathbf{y}_i + \mathbf{y}_j) - 2\mu|^{-1}\}}.$$

Consider next the limiting distribution of the HL estimate $\hat{\mu}(\mathbf{Y})$ under mild assumption.

Assumption 2 *The density of* $\mathbf{y}_i$ *is uniformly bounded and continuous with a unique spatial median minimizing*

$$D(\mu) = \mathbf{E}\left\{|\mathbf{y}_i + \mathbf{y}_j - 2\mu| - |\mathbf{y}_i + \mathbf{y}_j|\right\}, \quad i \neq j.$$

Again, for the asymptotical results, it is not a restriction to assume that $\mu = \mathbf{0}$. As in the case of the spatial median, both $D_n(\mu)$ and $D(\mu)$ are again convex and one can use Lemmas 6.1 and 6.3 for the pointwise results

$$D_n(\mu) \to_P D(\mu) = 4\mu'\mathbf{A}\mu + o(|\mu|^2)$$

and

$$nD_n(n^{-1/2}\mu) - 4(\sqrt{n}\mathbf{T} - \mathbf{A}\mu)'\mu = o_P(1),$$

where

$$\mathbf{A} = \mathbf{E}\left\{\mathbf{A}(\mathbf{y}_i + \mathbf{y}_j)\right\}, \quad i \neq j$$

with $\mathbf{A}(\mathbf{y}) = |\mathbf{y}|^{-1}(\mathbf{I} - |\mathbf{y}|^{-2}\mathbf{y}\mathbf{y}')$. Then Theorem B.5 in Appendix B gives the following result.

Theorem 7.4. *Under Assumption 2,* $\hat{\mu} \to_P \mu$ *and the limiting distribution of* $\sqrt{n}(\hat{\mu} - \mu)$ *is*

$$N_p\left(\mathbf{0}, \mathbf{A}^{-1}\mathbf{B}\mathbf{A}^{-1}\right).$$

We next define sample statistics needed in the estimation of the limiting covariance matrix of the HL estimate. Now write

$$\hat{\mathbf{A}} = \mathbf{A}(\mathbf{Y} - \mathbf{1}_n\hat{\mu}') = \mathbf{AVE}\left\{\mathbf{A}\left(\mathbf{y}_i + \mathbf{y}_j - 2\hat{\mu}\right)\right\} \quad \text{and} \quad \hat{\mathbf{B}} = \mathbf{B}(\mathbf{Y} - \mathbf{1}_n\hat{\mu}')$$

which, under the stated assumption, converge in probability to the population values

$$\mathbf{A} = \mathbf{E}\left\{\mathbf{A}\left(\varepsilon_i + \varepsilon_j\right)\right\}, \quad i \neq j, \quad \text{and} \quad \mathbf{B} = \mathbf{E}\left\{\mathbf{Q}_F(\varepsilon_i)\mathbf{Q}_F(\varepsilon_i)'\right\},$$

respectively. (The proof is similar to that in the spatial median case.) Theorem 7.4 suggests that the distribution of $\hat{\mu}(\mathbf{Y})$ can be approximated by

$$N_p\left(\mu, \frac{1}{n}\hat{\mathbf{A}}^{-1}\hat{\mathbf{B}}\hat{\mathbf{A}}^{-1}\right),$$

and approximate confidence ellipsoids for μ can be constructed.

The *spatial Hodges-Lehmann estimate with the inner standardization* is obtained by linking the spatial Hodges-Lehmaan estimate with the scatter matrix based on spatial signed-ranks in the following way.

Definition 7.4. Let μ be a p-vector and $\mathbf{S} > 0$ a symmetric $p \times p$ matrix, and define

$$\varepsilon_i = \varepsilon_i(\mu, \mathbf{S}) = \mathbf{S}^{-1/2}(\mathbf{y}_i - \mu), \quad i = 1, \ldots, n.$$

The simultaneous estimates of location and scatter based on the spatial signed-rank score function are the values of μ and $\mathbf{S}$ that satisfy

$$\text{AVE}\left\{\mathbf{Q}(\varepsilon_i)\right\} = \mathbf{0} \quad \text{and} \quad p\,\text{AVE}\left\{\mathbf{Q}(\varepsilon_i)\mathbf{Q}(\varepsilon_i)'\right\} = \text{AVE}\left\{\mathbf{Q}(\varepsilon_i)'\mathbf{Q}(\varepsilon_i)\right\}\mathbf{I}_p.$$

This pair of location and scatter estimates is easy to compute. Again the first iteration step updates the residuals, the second one updates the location center, and finally the third one updates the scatter matrix as follows.

1.
$$\varepsilon_i \leftarrow \mathbf{S}^{-1/2}(\mathbf{y}_i - \mu), \quad i = 1, \ldots, n.$$

2.
$$\mu \leftarrow \mu + \frac{1}{2}\frac{\mathbf{S}^{1/2}\,\mathbf{AVE}\{\mathbf{U}(\varepsilon_i + \varepsilon_j)\}}{\mathbf{AVE}\{|\varepsilon_i + \varepsilon_j|^{-1}\}}.$$

3.
$$\mathbf{S} \leftarrow p\,\mathbf{S}^{1/2}\,\mathbf{QCOV}(\varepsilon)\,\mathbf{S}^{1/2}.$$

Note, however, that the location estimate with inner standardization is now for the transformation-retransformation Hodges-Lehmann center, not for the regular Hodges-Lehmann center. In the symmetric case the centers of course are the same.

Example 7.2. **Cork boring data** Consider again the two datasets, one with a 3-variate vector of E-N, S-N and W-N and one with a bivariate S-N and W-E. We wish to estimate the Hodges-Lehmann estimates in the 3-variate case and in the bivariate cases. We give an R code to find the estimate and its covariance matrix. To compare the estimates, the 95% confidence ellipsoids for the three estimates are illustrated in Figures 7.3 and 7.4.

```
>% 3-variate case
> est <- mv.1sample.est(cork_3v)
> est.sign.i.3v <- mv.1sample.est(cork_3v, "s", "i")
> est.signrank.i.3v <- mv.1sample.est(cork_3v, "r", "i")
> summary(est.signrank.i.3v)
The equivariant spatial Hodges-Lehmann estimator of cork_3v is:
[1] -3.9246 -0.6865 -4.8635

And has the covariance matrix:
       [,1]   [,2]   [,3]
[1,] 2.0198 0.2974 0.6502
[2,] 0.2974 1.8848 1.1947
[3,] 0.6502 1.1947 2.5190
>
> plotMvloc(est, est.sign.i.3v, est.signrank.i.3v, X=cork_3v,
  alim="e", color.ell=1:3, , lty.ell=1:3, pch.ell= 15:17)
>
>% 2-variate case
> est <- mv.1sample.est(cork_2v)
> est.sign.i.2v <- mv.1sample.est(cork_2v, "s", "i")
> est.signrank.i.2v <- mv.1sample.est(cork_2v, "r", "i")
> summary(est.signrank.i.2v)
The equivariant spatial Hodges-Lehmann estimator of cork_2v is:
[1] -0.6854 -0.7337

And has the covariance matrix:
      [,1]  [,2]
[1,] 1.677 1.025
[2,] 1.025 3.198
>
> plotMvloc(est, est.sign.i.2v, est.signrank.i.2v, X=cork_2v,
  color.ell=1:3, , lty.ell=1:3, pch.ell= 15:17)
>
```

7.3 Other approaches

The signed-rank scores tests by Puri and Sen (1971) combine marginal signed-rank scores tests in the widest symmetric nonparametric model. These tests are not affine

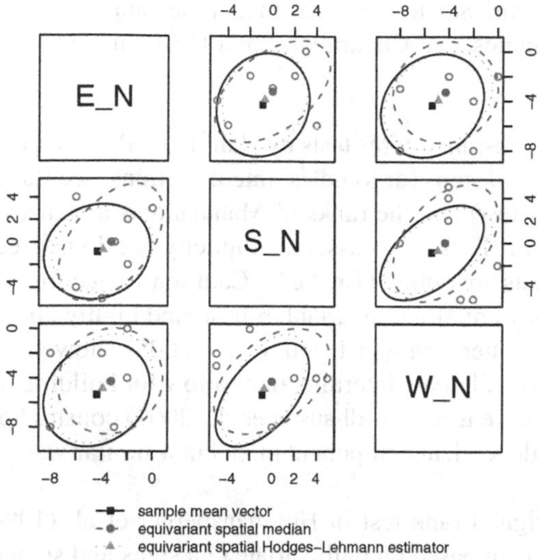

Fig. 7.3 The standardized spatial signed-ranks for the 3-variate data.

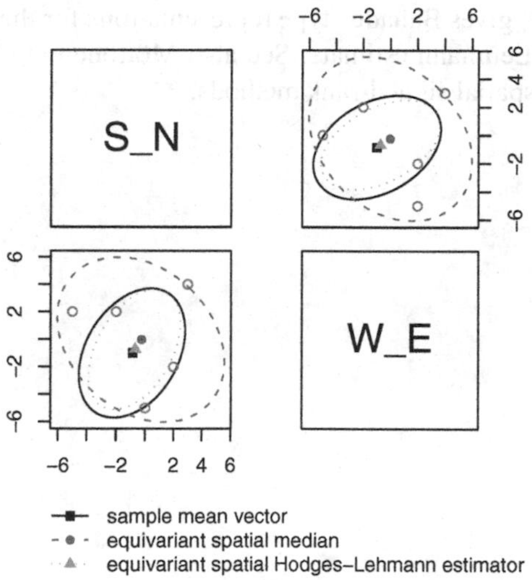

Fig. 7.4 The standardized spatial signed-ranks for the 2-variate data.

invariant and may have poor efficiency if the marginal variables are dependent. Invariant versions of Puri-Sen tests are obtained if the data points are first transformed to invariant coordinates; see Chakraborty and Chaudhuri (1999) and Nordhausen et al. (2006).

The optimal signed-rank scores tests by Hallin and Paindaveine (2002) are based on standardized spatial signs (or Randles' interdirections; see Randles (1989) for the corresponding sign test) and the ranks of Mahalanobis distances between the data points and the origin. These tests assume ellipticity but do not require any moment assumption. The tests are optimal (in the Le Cam sense) at correctly specified (elliptical) densities. They are affine-invariant, robust, and highly efficient under a broad range of densities. Later Oja and Paindaveine (2005) showed that interdirections together with the so-called lift-interdirections allow for building totally hyperplane-based versions of these tests. Nordhausen et al. (2009) constructed optimal signed-rank tests in the independent component model in a similar way.

The sign and signed-rank test in Hettmansperger et al. (1994) and Hettmansperger et al. (1997) are based on multivariate Oja signs and signed-ranks. They can be used in all models above, are asymptotically equivalent to spatial sign and signed-rank tests in the spherical case, and are affine-invariant. However, at the elliptic model, their efficiency (as well as that of the spatial sign and signed-rank tests) may be poor when compared with the Hallin and Paindaveine tests.

Chaudhuri (1992) gives Bahadur-type representations for the spatial median and the spatial Hodges-Lehmann estimate. See also Möttönen et al. (2005) for multivariate generalized spatial signed-rank methods.

Chapter 8
One-sample problem: Comparisons of tests and estimates

Abstract The efficiency and robustness properties of the tests and estimates are discussed. The estimates (the mean vector, the spatial median, and the spatial Hodges-Lehmann estimate) are compared using their limiting covariance matrices. The Pitman asymptotical relative efficiencies (ARE) of the spatial sign and spatial signed-rank tests with respect to the Hotelling's T^2 are considered in the multivariate t distribution case. The tests using inner and outer standardizations are compared as well. Simulation studies and some analyses of real datasets are used to illustrate the difference between the estimates and between the tests.

8.1 Asymptotic relative efficiencies

In this section we first consider the *limiting Pitman efficiences* of the spatial sign and signed-rank tests with respect to the classical Hotelling's T^2-test in the one-sample location case. In this comparison we assume that $\mathbf{Y}$ is a random sample from a symmetrical distribution around μ with a p-variate density function $f(\mathbf{y}-\mu)$. Here $f(\mathbf{y})$ is a density function symmetrical around the origin; that is, $f(-\mathbf{y}) = f(\mathbf{y})$. We wish to test the null hypothesis

$$H_0: \ \mu = \mathbf{0}.$$

We write $\mathbf{L}(\mathbf{y}) = -\nabla \log f(\mathbf{y})$ for the *optimal location score function.* We also assume that the Fisher information matrix $\mathscr{I} = \mathbf{E}\{\mathbf{L}(\mathbf{y}_i)\mathbf{L}(\mathbf{y}_i)'\}$ is bounded.

Consider the *score test statistics* of the general form

$$\mathbf{T} = \mathbf{T}(\mathbf{Y}) = \mathbf{AVE}\{\mathbf{T}(\mathbf{y}_i)\}$$

for a p-vector valued function $\mathbf{T}(\mathbf{y})$. In the one-sample location case, it is natural to assume that $\mathbf{T}$ is odd so that $\mathbf{E}\{\mathbf{T}(\mathbf{y}_i)\} = \mathbf{0}$. It is then well known that, if one is interested in the high efficiency of the test, the best choice for the score function

H. Oja, *Multivariate Nonparametric Methods with R: An Approach Based on Spatial Signs and Ranks*, Lecture Notes in Statistics 199, DOI 10.1007/978-1-4419-0468-3_8,

$\mathbf{T}(\mathbf{y})$ is the optimal location score function $\mathbf{L}(\mathbf{y})$. Write

$$\mathbf{L} = \mathbf{L}(\mathbf{Y}) = \mathbf{AVE}\{\mathbf{L}(\mathbf{y}_i)\}$$

for this optimal test statistic. Also recall that the identity score function, $\mathbf{T}(\mathbf{y}) = \mathbf{y}$, yields a test that is asymptotically equivalent with Hotelling's T^2 and optimal under multivariate normality.

Using the multivariate central limit theorem we get the followin lemma.

Lemma 8.1. *Assume that*

$$\mathbf{A} = \mathbf{E}\{\mathbf{T}(\mathbf{y}_i)\mathbf{L}(\mathbf{y}_i)'\} \quad \textit{and} \quad \mathbf{B} = \mathbf{E}\{\mathbf{T}(\mathbf{y}_i)\mathbf{T}(\mathbf{y}_i)'\}$$

exist and are bounded. Then

$$n^{1/2}\begin{pmatrix}\mathbf{T}\\ \mathbf{L}\end{pmatrix} \to_d N_{2p}\left(\begin{pmatrix}\mathbf{0}\\ \mathbf{0}\end{pmatrix}, \begin{pmatrix}\mathbf{B} & \mathbf{A}\\ \mathbf{A}' & \mathscr{I}\end{pmatrix}\right).$$

Theorem 8.1. *Assume that the alternative sequences of the form $H_n: \ f(\mathbf{y} - n^{-1/2}\delta)$ are contiguous satisfying, under H_0,*

$$\sum_{i=1}^{n} \log\left(\frac{f(\mathbf{y}_i - n^{-1/2}\delta)}{f(\mathbf{y}_i)}\right) = n^{1/2}\mathbf{L}'\delta - \frac{1}{2}\delta'\mathscr{I}\delta + o_P(1).$$

Then, under the alternative sequences H_n, the limiting distribution of the test statistic $n^{1/2}\mathbf{T}$ is a p-variate normal distribution with mean vector $\mathbf{A}\delta$ and covariance matrix $\mathbf{B}$.

Proof. See Möttönen et al. (1997).

Corollary 8.1. *Under the sequence of contiguous alternatives the limiting distribution of the squared test statistic $Q^2 = n\mathbf{T}'\mathbf{B}^{-1}\mathbf{T}$ is a noncentral chi-square distribution with p degrees of freedom and the noncentrality parameter $\delta'\mathbf{A}'\mathbf{B}^{-1}\mathbf{A}\delta$.*

In the case of the null distribution, we could show (using Slutsky's theorem) that

$$n\mathbf{T}'\hat{\mathbf{B}}^{-1}\mathbf{T} - n\mathbf{T}'\mathbf{B}^{-1}\mathbf{T} \to_P 0.$$

This is then true for the contiguous alternative sequences as well, and the limiting χ_p^2 distribution with noncentrality parameter $\delta'\mathbf{A}'\mathbf{B}^{-1}\mathbf{A}\delta$ holds also for the tests where the true value $\mathbf{B}$ is replaced by its convergent estimate $\hat{\mathbf{B}}$.

As all the test statistics have limiting distributions of the same type, χ^2_p, the Pitman *asymptotical relative efficiencies (ARE)* of the multivariate sign test and multivariate signed-rank test relative to Hotelling's T^2, are simply the ratios of the noncentrality parameters,

$$ARE = \frac{\delta' \mathbf{A}' \mathbf{B}^{-1} \mathbf{A} \delta}{\delta' \Sigma^{-1} \delta}.$$

In the following we compare the efficiency of the tests in the case of spherically symmetrical distribution F. Assume that F is the distribution of $\mathbf{y}$ and write, as before, $r = |\mathbf{y}|$ and $\mathbf{u} = |\mathbf{y}|^{-1}\mathbf{y}$. In the spherically symmetric case the Pitman ARE of the spatial sign test with respect to the Hotelling's T^2 is then simply

$$ARE_1 = \left(\frac{p-1}{p}\right) \mathbf{E}\{r^2\} \mathbf{E}^2\{r^{-1}\}.$$

The asymptotic relative efficiencies when the underlying population is a multivariate spherical t distribution were derived by Möttönen et al. (1997). The optimal score function of the p-variate t-distribution with ν degrees of freedom, $t_{\nu,p}$, is

$$\mathbf{L}(\mathbf{y}) = \frac{\nu + p}{\nu + \mathbf{y}'\mathbf{y}} \mathbf{y}.$$

The theoretical signed-rank function $\mathbf{Q}(\mathbf{y}) = q(r)\mathbf{u}$ in the multivariate normal and in the t distribution cases is given in Examples 4.3 and 4.4. The Pitman efficiency of the spatial signed-rank test with respect to Hotelling's test is then

$$ARE_2 = \frac{\nu(\nu+p)^2}{p(\nu-2)} \mathbf{E}^2 \left\{ q(r) \frac{r}{\nu + r^2} \right\} \left[\mathbf{E}\{q^2(r)\} \right]^{-1},$$

where r^2/p has an $F(p,\nu)$ distribution. If the observations come from a multivariate normal distribution $N_p(\mathbf{0}, \mathbf{I})$, then we get

$$ARE_2 = \frac{1}{p} \mathbf{E}^2\{q(r)r\} [\mathbf{E}\{q^2(r)\}]^{-1},$$

where now r^2 has a χ^2_p distribution. See Möttönen et al. (1997).

The efficiencies in the multivariate t distribution case with some choices of ν are displayed in Table 8.1. We see that as the dimension p increases and as the distribution gets heavier tailed (ν gets smaller), the performances of the spatial sign test and the spatial signed-rank test improve relative to T^2. The sign test and the signed-rank test are clearly better than T^2 in heavy-tailed cases. For high dimensions and very heavy tails, the sign test is the most efficient one. Note that $\nu = \infty$ is the multivariate normal case.

In the comparison of the asymptotical efficiencies of the estimates of the location center μ, we first note that all the estimates $\hat{\mu}$ considered here (the mean vector, the

Table 8.1 Asymptotic relative efficiencies of the sign test and the signed-rank test relative to Hotelling's T^2 under p-variate t distributions with ν degrees of freedom for selected values of p and ν.

	Sign test			Signed-rank test		
dimension p	$\nu=3$	$\nu=6$	$\nu=\infty$	$\nu=3$	$\nu=6$	$\nu=\infty$
1	1.62	0.88	0.64	1.90	1.16	0.95
2	2.00	1.08	0.78	1.95	1.19	0.97
4	2.25	1.22	0.88	2.02	1.21	0.98
10	2.42	1.31	0.95	2.09	1.22	0.99

spatial median and the spatial Hodges-Lehmann estimate) are root-n consistent and

$$\sqrt{n}(\hat{\mu}-\mu) \to_D N_p(\mathbf{0}, \mathbf{A}^{-1}\mathbf{B}\mathbf{A}^{-1}),$$

where, as before,

$$\mathbf{A} = \mathbf{E}\{\mathbf{T}(\mathbf{y}_i)\mathbf{L}(\mathbf{y}_i)'\} \quad \text{and} \quad \mathbf{B} = \mathbf{E}\{\mathbf{T}(\mathbf{y}_i)\mathbf{T}(\mathbf{y}_i)'\}$$

depend on the chosen score $\mathbf{T}(\mathbf{y})$ and the distribution F. The comparison of the estimates is then based on the asymptotic covariance matrix $\mathbf{A}^{-1}\mathbf{B}\mathbf{A}^{-1}$ and possible global measures of variation are, for example, the geometric mean or the arithmetic mean of the eigenvalues, that is,

$$\left(\det(\mathbf{A}^{-1}\mathbf{B}\mathbf{A}^{-1})\right)^{1/p} \quad \text{or} \quad tr\left(\mathbf{A}^{-1}\mathbf{B}\mathbf{A}^{-1}\right)/p.$$

The former is more natural as it is invariant under affine transformations to the original observations. Note also that the volume of the approximate confidence ellipsoid is proportional to $\det(\mathbf{A}^{-1}\mathbf{B}\mathbf{A}^{-1})$. In the case of elliptically symmetric distributions, it is then enough to consider the spherical cases only and, when comparing two estimates, the ratio of the geometric means of the eigenvalues is the same as the asymptotic relative efficiency of the corresponding tests. The asymptotical efficiencies listed in Table 8.1, for example, then also hold true for the corresponding estimates.

8.2 Finite sample comparisons

Example 8.1. **Estimates and confidence ellipsoids for data sets with outliers.** In our first example we consider again the 3-variate and bivariate datasets with measurements E-N, S-N and W-N and S-N and W-E, respectively. We use the estimates with inner standardization in the comparisons. Figure 8.1 shows how different estimates and corresponding confidence ellipsoids change if the first observation in the 3-variate data set is moved to $(-50, 50, -50)'$ (one outlier). The estimates and the confidence ellipsoids for the original dataset are given in Figure 7.3. Note that the sample mean and the corresponding confidence ellipsoid react strongly to the outlying observation. The sample mean moves in the direction of the outlier and the shape of the ellipsoid is also changing. The R code used for the comparison is as follows.

```
> cork_3v_cont <- cork_3v
> cork_3v_cont[1,] <- cork_3v[1,] + c(-50, 50, -50)
>
> est.c <- mv.1sample.est(cork_3v_cont)
> est.sign.i.3v.c <- mv.1sample.est(cork_3v_cont, "s", "i")
> est.signrank.i.3v.c <- mv.1sample.est(cork_3v_cont, "r", "i")
>
> plotMvloc(est.c, est.sign.i.3v.c, est.signrank.i.3v.c,
  X=cork_3v_cont,
  alim="e", color.ell=1:3, , lty.ell=1:3, pch.ell= 15:17)
```

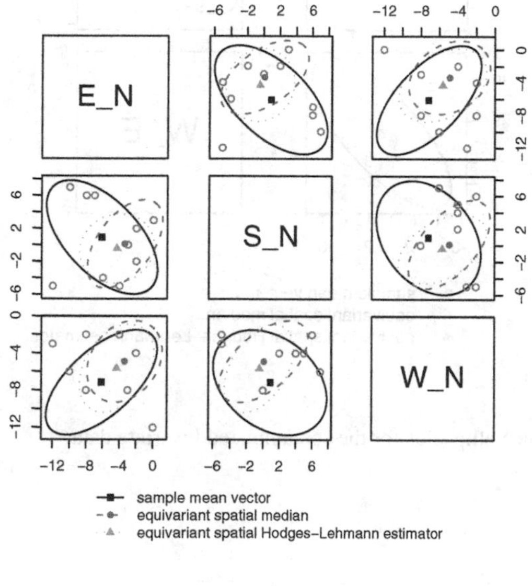

Fig. 8.1 The confidence ellipsoids for the contaminated 3-variate data.

In Figure 8.2 a similar behavior of the estimates is illustrated for the bivariate data. Now the first observation is replaced by $(50,50)'$. The estimates and the confidence ellipsoids for the original data set are given in Figure 7.4. The R code follows.

```
> cork_2v_cont <- cork_2v
> cork_2v_cont[1,] <- cork_2v[1,] + c(50, 50)
>
> est.c <- mv.1sample.est(cork_2v_cont)
> est.sign.i.2v.c <- mv.1sample.est(cork_2v_cont, "s", "i")
> est.signrank.i.2v.c <- mv.1sample.est(cork_2v_cont, "r", "i")
>
> plotMvloc(est.c, est.sign.i.2v.c, est.signrank.i.2v.c,
  X=cork_2v_cont,
  alim="e", color.ell=1:3, , lty.ell=1:3, pch.ell= 15:17)
```

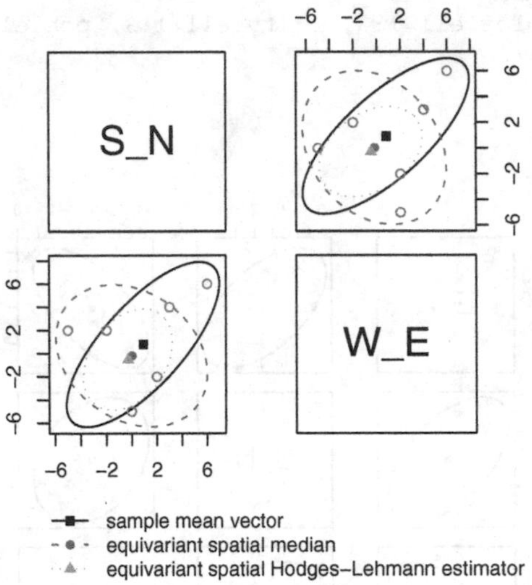

Fig. 8.2 The confidence ellipsoids for the contaminated bivariate data.

Example 8.2. **Comparison of the estimates in the multivariate normal and t distribution cases.** To illustrate the finite sample efficiencies of the estimates we generated random samples of size $n = 100$ from a bivariate normal distribution and from a 3-variate t_3 distribution. The estimates with corresponding 95 % confidence ellipsoids are found in Figures 8.3 and 8.4.

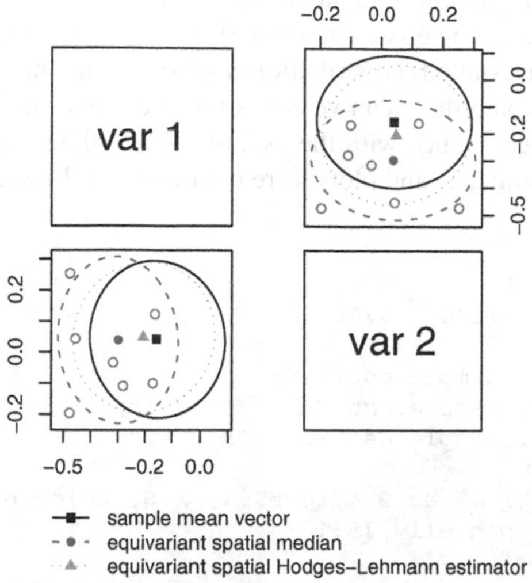

Fig. 8.3 A sample from a bivariate normal distribution: estimates with 95% confidence ellipsoids.

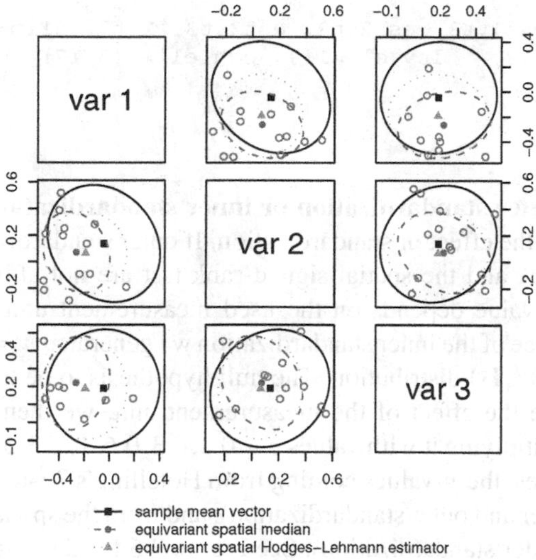

Fig. 8.4 A sample from a 3-variate t_3 distribution: estimates with 95% confidence ellipsoids.

As expected, in the multivariate normal case, the accuracies of the sample mean vector and the Hodges-Lehmann estimate are almost the same (the asymptotic relative efficiency of the HL-estimate is close to one), and the spatial median is poorest in this sense. In the heavy-tailed t_3 distribution case, the spatial median has the smallest confidence ellipsoid, and the mean vector is now very poor in its efficiency. These results are well in accordance with the asymptotical efficiencies reported in Table 8.1. The datasets, estimates, and plots were obtained as follows.

```
> set.seed(1234)
> X.N <- rmvnorm(100,c(0,0))
>
> est1.N <- mv.1sample.est(X.N)
> est2.N <- mv.1sample.est(X.N, "s", "i")
> est3.N <- mv.1sample.est(X.N, "r", "i")
>
> plotMvloc(est1.N, est2.N, est3.N, X.N, color.ell=1:3, ,
  lty.ell=1:3, pch.ell= 15:17)
>
> set.seed(1234)
> X.t3 <- rmvt(100, diag(3), 3)
>
> est1.t3 <- mv.1sample.est(X.t3)
> est2.t3 <- mv.1sample.est(X.t3, "s", "i")
> est3.t3 <- mv.1sample.est(X.t3, "r", "i")
>
> plotMvloc(est1.t3, est2.t3, est3.t3, X.t3, alim="e",
  color.ell=1:3, , lty.ell=1:3, pch.ell= 15:17)
```

Example 8.3. **Outer standardization or inner standardization?** In this small study we consider the effect of standardization. If outer standardization is used then the spatial sign test and the spatial signed-rank test are not affine invariant. This means that the p-value depends on the used measurement unit, for example. To show the importance of the inner standardization we generated $n = 150$ observations from an $N_p((0.1,0)', \mathbf{I}_2)$ distribution. The null hypothesis to be tested, $H_0: \mu = \mathbf{0}$, is not true. To see the effect of the measurement unit, we then rescaled the first component by multiplying it with values $c = 0.1, 0.3, 0.6, 1, 2, 3, 5$ and calculated, in all these seven cases, the p-values coming from Hotelling's T^2-test, from the spatial sign test (with inner and outer standardization), and from the spatial signed-rank test (with inner and outer standardization). See Figure 8.5 for the results. The p-values of the spatial sign test and the spatial signed-rank test with outer standardization depend strongly on the rescaling constant c. Note that the p-value obtained from the spatial sign test with outer standardization varies from the smallest p-value to the largest one with c. The invariant test versions behave as expected: the p-values are in the order of the asymptotic efficiency of the tests. The R code used is lengthy but available on request.

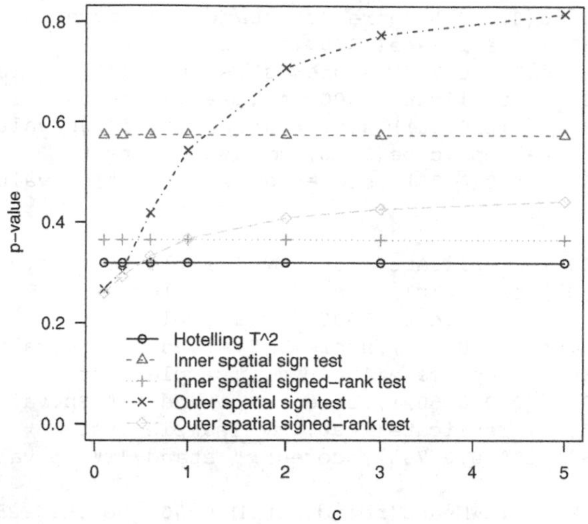

Fig. 8.5 The p-values of the tests as a function of the measurement unit. The second component is multiplied by c.

Example 8.4. **Simulation studies to compare the tests.** Next we compare the finite sample efficiencies of the three competing tests, Hotelling's T^2-test, the spatial sign test, and the spatial signed-rank test. Again the samples of sizes $n = 50$ were simulated from a 3-variate standard normal distribution and from a 3-variate t_3 distribution with covariance matrices $\mathbf{I}_3$. The powers of the tests for alternatives $\mu = (0,0,0)'$, $(0,0,0.25)'$, $(0,0,0.50)'$, and $(0,0,0.75)'$ were estimated by generating 1000 samples of size $n = 50$ from all these distributions. In the tests we used the asymptotical critical value $\chi^2_{p,0.95}$. The R code for the simulation in the 3-variate normal case with the results follows.

```
> # Simulation in the normal case
>
> set.seed(1)
> Hot.N.0.00 <- replicate(1000, mv.1sample.test
  (rmvnorm(50, c(0,0,0)))$p.value)
> Hot.N.0.25 <- replicate(1000, mv.1sample.test
  (rmvnorm(50, c(0,0,0.25)))$p.value)
> Hot.N.0.50 <- replicate(1000, mv.1sample.test
  (rmvnorm(50, c(0,0,0.50)))$p.value)
> Hot.N.0.75 <- replicate(1000, mv.1sample.test
  (rmvnorm(50, c(0,0,0.75)))$p.value)
>
```

```
> set.seed(1)
> Sign.N.0.00 <- replicate(1000, mv.1sample.test
 (rmvnorm(50, c(0,0,0)),score="s",stand="i")$p.value)
> Sign.N.0.25 <- replicate(1000, mv.1sample.test
 (rmvnorm(50, c(0,0,0.25)),score="s",stand="i")$p.value)
> Sign.N.0.50 <- replicate(1000, mv.1sample.test
 (rmvnorm(50, c(0,0,0.50)),score="s",stand="i")$p.value)
> Sign.N.0.75 <- replicate(1000, mv.1sample.test
 (rmvnorm(50, c(0,0,0.75)),score="s",stand="i")$p.value)
>
> set.seed(1)
> Rank.N.0.00 <- replicate(1000, mv.1sample.test(rmvnorm
 (50, c(0,0,0)),score="r",stand="i")$p.value)
> Rank.N.0.25 <- replicate(1000, mv.1sample.test
 (rmvnorm(50, c(0,0,0.25)),score="r",stand="i")$p.value)
> Rank.N.0.50 <- replicate(1000, mv.1sample.test
 (rmvnorm(50, c(0,0,0.50)),score="r",stand="i")$p.value)
> Rank.N.0.75 <- replicate(1000, mv.1sample.test
  (rmvnorm(50, c(0,0,0.75)),score="r",stand="i")$p.value)
>
> power.Hot.N <- rowMeans(rbind(Hot.N.0.00, Hot.N.0.25,
  Hot.N.0.50, Hot.N.0.75) <= 0.05)
> power.Sign.N <- rowMeans(rbind(Sign.N.0.00, Sign.N.0.25,
  Sign.N.0.50, Sign.N.0.75) <= 0.05)
> power.Rank.N <- rowMeans(rbind(Rank.N.0.00, Rank.N.0.25,
  Rank.N.0.50, Rank.N.0.75) <= 0.05)
>
> res.N <- cbind(delta= seq(0,0.75,0.25), power.Hot.N,
  power.Sign.N, power.Rank.N)
> rownames(res.N) <- NULL
> res.N
     delta power.Hot.N power.Sign.N power.Rank.N
[1,]  0.00       0.058        0.040        0.035
[2,]  0.25       0.323        0.235        0.251
[3,]  0.50       0.873        0.747        0.818
[4,]  0.75       0.996        0.984        0.992
>
```

First note that the spatial sign test and the spatial signed-rank test seem too conservative; the true rejection probability seems to be smaller than 0.05. Therefore the permutation version of the tests should be used for small sample sizes. The Hotelling test is naturally the best one in this case. Next the results in the t_3 distribution case follow. Again, the spatial sign test is most efficient in this case. This is in agreement with the asymptotical relative efficiencies of the tests.

```
> res.T
     delta power.Hot.T power.Sign.T power.Rank.T
[1,]  0.00       0.050        0.039        0.031
[2,]  0.25       0.190        0.215        0.204
```

```
[3,]  0.50          0.511          0.651          0.581
[4,]  0.75          0.835          0.958          0.930
>
```

Chapter 9
One-sample problem: Inference for shape

Abstract The one-sample multivariate shape problem is considered. The shape matrix is a scatter matrix rescaled or normalized in a certain way. The procedures here are based on multivariate spatial signs and spatial ranks. Tests and estimates based on the spatial sign covariance matrix and the spatial rank covariance matrices of different types are considered. The asymptotic efficiencies show a good performance of these methods, particularly for heavy-tailed distributions.

9.1 The estimation and testing problem

We consider the one-sample case where $\mathbf{Y} = (\mathbf{y}_1, \mathbf{y}_2, ..., \mathbf{y}_n)'$ is a random sample from a symmetrical distribution. We again assume that the observations are generated by

$$\mathbf{y}_i = \mu + \Omega \varepsilon_i, \quad i = 1, ..., n,$$

where the ε_i are centered and standardized residuals with cumulative distribution function $F(\varepsilon)$ and density function $f(\varepsilon)$. Different assumptions on the distribution of ε_i yield different parametric, nonparametric, and semiparametric models. Parameter μ is the unknown location center, and $\Sigma = \Omega\Omega'$ the unknown scatter matrix.

In previous chapters we tested the null hypothesis $H_0: \ \mu = \mathbf{0}$ and estimated unknown μ. The tests and estimates were based on different score functions (identity, spatial sign, and spatial signed-rank). The inner standardization to attain the affine invariance of the location tests or the affine equivariance of the location estimates also yielded the corresponding scatter matrix or shape matrix estimates. In this chapter we compare the shape matrix estimates and the corresponding tests obtained in this way. The proofs of the results here can mostly be found in Sirkiä et al. (2008).

H. Oja, *Multivariate Nonparametric Methods with R: An Approach Based on Spatial Signs and Ranks*, Lecture Notes in Statistics 199, DOI 10.1007/978-1-4419-0468-3_9,

We recall that assumptions on the distribution of ε_i were needed to fix parameters μ and Σ in a natural way. In all symmetric models (A0)–(A4) parameter μ is well defined as the unique symmetry center of the distribution of $\mathbf{y}_i$. Scatter parameter Σ is uniformly defined only in the multivariate normal model (A0), and defined up to a constant in the elliptic model (A1) as well as in the location-scatter model (A2). We refer back to Chapter 2 for a discussion of the symmetric models.

In this chapter we consider the inference tools for Σ in models (A0), (A1), and (A2), and the main focus is on the procedures based on the spatial signs and ranks. The *scatter parameter* Σ is assumed to be nonsingular and it is decomposed, as in Section 2.2, into two parts by $\Sigma = \sigma^2 \Lambda$ where $\sigma^2 = \sigma^2(\Sigma) > 0$ is a scalar-valued *scale parameter* and $\Lambda = \sigma^{-2}\Sigma$ is a matrix-valued *shape parameter*. The scale functional $\sigma^2(\Sigma)$ is supposed to satisfy

$$\sigma^2(\mathbf{I}_p) = 1 \quad \text{and} \quad \sigma^2(c\Sigma) = c\sigma^2(\Sigma).$$

In the literature, the scale $\sigma^2 = \sigma^2(\Sigma)$ has been defined as

$$\Sigma_{11}, \quad \frac{tr(\Sigma)}{p}, \quad \frac{p}{tr(\Sigma^{-1})}, \quad \text{or} \quad \det(\Sigma)^{1/p}.$$

The shape matrix Λ can be seen as a *normalized version* of the scatter matrix Σ and is a well-defined parameter in models (A0), (A1), and (A2). See also Paindaveine (2008). We wish to test the null hypothesis of sphericity, $H_0 : \Lambda = \mathbf{I}_p$, and estimate the unknown value of Λ. This is not a restriction: if one is interested in testing the null hypothesis $H_0 : \Lambda = \Lambda_0$, then one can first transform $\mathbf{y}_i \to \Lambda_0^{-1/2}\mathbf{y}_i$ and then apply the test to the transformed observations.

In many classical problems in multivariate analysis it is sufficient to base the analysis on the estimate of Λ only. The applications include the (robust) estimation of the correlation matrix, principal component analysis (PCA), canonical correlation analysis (CCA), and multivariate regression analysis among others.

Most of the results here are given under the assumption of elliptical symmetry. Recall that a random variable $\mathbf{y}_i$ is elliptically symmetric if ε_i is spherically symmetric. The density function of $\mathbf{y}_i$ is then of the form

$$\det(\Sigma)^{-1/2} f\left(\Sigma^{-1/2}(\mathbf{y}-\mu)\right),$$

where

$$f(\varepsilon) = \exp(-\rho(|\varepsilon|))$$

with some function ρ. Then parameter μ is the symmetry center of the distribution of $\mathbf{y}_i$, and parameter Σ is a positive definite symmetric $p \times p$ scatter matrix. It is easy to see that the shape matrix estimate

$$\hat{\Lambda} = \sigma^2(\mathbf{S})^{-1}\mathbf{S}$$

based on any consistent scatter matrix $\mathbf{S}$ is a consistent estimate of the corresponding population quantity Λ.

9.2 Important matrix tools

Here we recall some matrix tools introduced and already used in Section 3.3. See also Appendix A. As before, $\mathbf{K}_{p,p}$ is the commutation matrix, that is, a $p^2 \times p^2$ block matrix with (i,j)-block being equal to a $p \times p$ matrix that has one at entry (j,i) and zero elsewhere, and $\mathbf{J}_{p,p}$ for $\text{vec}(\mathbf{I}_p)\text{vec}(\mathbf{I}_p)'$. These matrices $\mathbf{K}_{p,p}$ and $\mathbf{J}_{p,p}$ have the following interesting properties.

$$\mathbf{K}_{p,p}\text{vec}(\mathbf{A}) = \text{vec}(\mathbf{A}') \quad \text{and} \quad \mathbf{J}_{p,p}\text{vec}(\mathbf{A}) = tr(\mathbf{A})\text{vec}(\mathbf{I}_p).$$

Matrix

$$\mathbf{C}_{p,p} = \frac{1}{2}(\mathbf{I}_{p^2} + \mathbf{K}_{p,p}) - \frac{1}{p}\mathbf{J}_{p,p}$$

projects a vectorized matrix $\text{vec}(\mathbf{A})$ to the space of symmetrical and centered vectorized matrices. (Recall that $\mathbf{C}_{p,p} = \mathbf{P}_1 + \mathbf{P}_2$ where $\mathbf{P}_1$ and $\mathbf{P}_2$ are projections introduced in Section 3.3.) The tests and estimates for the shape parameter are based on the squared norm of such a projection,

$$Q^2(\mathbf{A}) = |\mathbf{C}_{p,p}\text{vec}(\mathbf{A})|^2,$$

which is proportional to the variance of the eigenvalues of a symmetrized version of $\mathbf{A}$. It is easy to see that, for symmetrical positive definite $p \times p$ matrices $\mathbf{A}$,

$$Q^2(\mathbf{A}) = 0 \quad \Leftrightarrow \quad \mathbf{A} \propto \mathbf{I}_p.$$

In the following, we also often need the notation

$$\mathbf{C}_{p,p}(\mathbf{V}) = \frac{1}{2}(\mathbf{I}_{p^2} + \mathbf{K}_{p,p})(\mathbf{V} \otimes \mathbf{V}) - \frac{1}{p}\text{vec}(\mathbf{V})\text{vec}(\mathbf{V})'.$$

Clearly $\mathbf{C}_{p,p}(\mathbf{I}_p) = \mathbf{C}_{p,p}$.

Finally, if $\mathbf{A}$ is a nonnegative definite symmetric $p \times p$ matrix with r positive eigenvalues $\lambda_1 \geq \cdots \geq \lambda_r > 0$, $p \geq r \geq 1$, and if $\mathbf{o}_1, ..., \mathbf{o}_r$ are the corresponding eigenvectors, then the Moore-Penrose inverse of the matrix

$$\mathbf{A} = \sum_{i=1}^{r} \lambda_i \mathbf{o}_i \mathbf{o}_i'$$

is

$$\mathbf{A}^- = \sum_{i=1}^{r} \lambda_i^{-1} \mathbf{o}_i \mathbf{o}_i'.$$

9.3 The general strategy for estimation and testing

A general idea to construct tests and estimates for location was to use a p-vector valued score function $\mathbf{T}(\mathbf{y})$ yielding individual scores $\mathbf{T}_i = \mathbf{T}(\mathbf{y}_i)$, $i = 1,...,n$. To attain affine equivariance/invariance of the location procedures we used either

- *Inner standardization of the scores*: Find transformation matrix $\mathbf{S}^{-1/2}$ such that, if $\hat{\mathbf{T}}_i = \mathbf{T}(\mathbf{S}^{-1/2}\mathbf{y}_i)$, then

$$p \cdot \mathbf{AVE}\{\hat{\mathbf{T}}_i \hat{\mathbf{T}}_i'\} = \mathbf{AVE}\{\hat{\mathbf{T}}_i' \hat{\mathbf{T}}_i\} \mathbf{I}_p;$$

or

- *Inner centering and standardization of the scores:* Find shift vector $\hat{\mu}$ and transformation matrix $\mathbf{S}^{-1/2}$ such that, if $\hat{\mathbf{T}}_i = \mathbf{T}(\mathbf{S}^{-1/2}(\mathbf{y}_i - \hat{\mu}))$, then

$$\mathbf{AVE}\{\hat{\mathbf{T}}_i\} = \mathbf{0} \quad \text{and} \quad p \cdot \mathbf{AVE}\{\hat{\mathbf{T}}_i \hat{\mathbf{T}}_i'\} = \mathbf{AVE}\{\hat{\mathbf{T}}_i' \hat{\mathbf{T}}_i\} \mathbf{I}_p.$$

In the first case, depending on the chosen score, one gets a scatter or shape matrix estimate $\mathbf{S} = \mathbf{S}(\mathbf{Y})$ with respect to the origin. In the second case, simultaneous estimates of location and scatter, $\hat{\mu}$ and $\mathbf{S}$, are obtained. Of course one should check separately in each case whether the estimates really exist for a data set at hand. Also it is not at all clear whether the estimates using different scores estimate the same population quantity. This did not present a problem in the location problem as the transformation $\mathbf{S}^{-1/2}$ was seen there only as a natural tool to attain affine invariance/equivariance.

The algorithm for the shape matrix estimate $\mathbf{S} = \mathbf{S}(\mathbf{Y})$ with respect to the origin then uses the following two steps.

1.
$$\hat{\mathbf{T}}_i \leftarrow \mathbf{T}(\mathbf{S}^{-1/2}\mathbf{y}_i), \quad i = 1,...,n; \qquad \hat{\mathbf{T}} \leftarrow (\hat{\mathbf{T}}_1,...,\hat{\mathbf{T}}_n)'.$$

2.
$$\mathbf{S} \leftarrow \frac{p}{tr(\hat{\mathbf{T}}'\hat{\mathbf{T}})} \mathbf{S}^{1/2} \hat{\mathbf{T}}' \hat{\mathbf{T}} \mathbf{S}^{1/2}.$$

The test statistic for testing $H_0 : \Lambda = \mathbf{I}_p$ is simply

$$Q^2\left(n^{-1}\mathbf{T}'\mathbf{T}\right) = \left|\mathbf{C}_{p,p}\text{vec}\left(n^{-1}\mathbf{T}'\mathbf{T}\right)\right|^2.$$

Note that in our approach $n^{-1}\mathbf{T}'\mathbf{T}$ is **COV**, **UCOV**, **TCOV** or **RCOV** depending on which score function is chosen.

Several approaches based on the regular covariance matrix for testing the shape in the multivariate normal and elliptic case can be found in the literature. These tests are thus based on the identity score $\mathbf{T}(\mathbf{y}) = \mathbf{y}$ and on the regular covariance matrix. Mauchly (1940) showed that in the multivariate normal distribution case the likelihood ratio test for testing the sphericity, that is, the null hypotheses $H_0 : \Lambda = \mathbf{I}_p$, is given by

$$L = \left\{ \frac{\det(\mathbf{COV})}{[tr(\mathbf{COV})/p]^p} \right\}^{n/2},$$

where $\mathbf{COV} = \mathbf{COV}(\mathbf{Y})$ is the regular sample covariance matrix. Note that L is essentially the ratio of two scale parameters $\det(\mathbf{COV})^{1/p}$ and $tr(\mathbf{COV})/p$. Under the null hypotheses, $-2\log L \sim \chi^2_{(p+2)(p-1)/2}$. Muirhead and Waternaux (1980) showed that the test based on L may also be used to test the sphericity under elliptical models with finite fourth moments. Later, Tyler (1983) obtained a robust version of the likelihood ratio test by replacing the sample covariance matrix with a robust scatter matrix estimator.

Also John (1971) and John (1972) considered the testing problem at the normal distribution case. He showed that the test

$$Q_J^2 = \frac{np^2}{2} \left| \frac{\mathbf{COV}}{tr(\mathbf{COV})} - \frac{1}{p} I_p \right|^2 = \frac{np^2}{2} Q^2 \left(\frac{\mathbf{COV}}{tr(\mathbf{COV})} \right),$$

is the locally most powerful invariant test for sphericity under the multivariate normality assumption. This test is, however, valid only under the multivariate normality assumption. In the wider elliptical model one can use a slight modification of John's test, which remains asymptotically valid under elliptical distributions but of course needs the assumption on the finite fourth moments. The modified John's test is defined as

$$Q_{J'}^2 = \frac{np^2}{2(1+\kappa_F)} Q^2 \left(\frac{\mathbf{COV}}{tr(\mathbf{COV})} \right),$$

where κ_F is the value of the classical kurtosis measure based on the standardized fourth moment of the marginal distribution, that is,

$$\kappa_F = \frac{E(\varepsilon_{ij}^4)}{E^2(\varepsilon_{ij}^2)} - 3.$$

In the multivariate normal case $\kappa_F = 0$. In practice, κ_F must be replaced by its estimate, the corresponding sample statistic. See also Muirhead and Waternaux (1980) and Tyler (1982).

9.4 Test and estimate based on UCOV

Assume that $\mathbf{Y} = (\mathbf{y}_1, ..., \mathbf{y}_n)$ is a random sample from an elliptically symmetric distribution with symmetry center $\mu = \mathbf{0}$ and shape parameter Λ. We wish to test the null hypothesis

$$H_0: \ \Lambda = \mathbf{I}_p.$$

The null hypothesis then says that the observations are coming from a spherical distribution. In efficiency studies we consider contiguous alternative sequences

$$H_n: \ \Lambda \propto \mathbf{I}_p + n^{-1/2}\mathbf{D},$$

where $\mathbf{D}$ is a symmetric matrix. Note that $\mathbf{D}$ fixes the "direction" for the alternative sequence.

The spatial sign covariance matrix is defined as

$$\mathbf{UCOV} = \mathbf{AVE}\left\{\mathbf{U}_i\mathbf{U}_i'\right\}.$$

Under the null hypothesis

$$\mathbf{E}(\text{vec}(\mathbf{UCOV})) = \frac{1}{p}\text{vec}(\mathbf{I}_p) \quad \text{and} \quad \mathbf{COV}(\text{vec}(\mathbf{UCOV})) = \frac{\tau}{n}\mathbf{C}_{p,p},$$

where

$$\tau = \frac{2}{p(p+2)},$$

and therefore also

$$\mathbf{E}(\mathbf{C}_{p,p}\text{vec}(\mathbf{UCOV})) = \mathbf{0} \quad \text{and} \quad \mathbf{COV}(\mathbf{C}_{p,p}\text{vec}(\mathbf{UCOV})) = \frac{\tau}{n}\mathbf{C}_{p,p}.$$

($\mathbf{C}_{p,p}$ is a projection matrix.)

The test statistic is proportional to the variance of the eigenvalues of the spatial sign covariance matrix.

Definition 9.1. The spatial sign test statistic is defined as

$$Q^2 = Q^2(\mathbf{UCOV}) = |\mathbf{C}_{p,p}\mathbf{UCOV}|^2.$$

Recall that the value of $Q^2(\mathbf{S})$ is equal to zero if and only if $\mathbf{S} \propto \mathbf{I}_p$. It is remarkable that the finite sample (and limiting) null distribution of Q^2 is the same for all spherical distributions as the sign covariance matrix $\mathbf{UCOV}$ depends on the observations only through their direction vectors. The limiting distribution of Q^2 under the null hypothesis as well as under the alternative sequence is given by the following.

Theorem 9.1. *Under the alternative sequence H_n,*

$$\frac{n}{\tau}Q^2 \to_d \chi^2_{(p+2)(p-1)/2}\left(\frac{1}{\tau(p+2)^2}Q^2(\mathbf{D})\right).$$

In a wider model (A3) where the shape parameter Λ is still welldefined, the covariance matrix of vec(**UCOV**) can be estimated by

$$\widehat{\mathbf{COV}}(\mathrm{vec}(\mathbf{UCOV})) = \frac{1}{n}\left[\mathbf{AVE}\left\{\mathbf{U}_i\mathbf{U}_i' \otimes \mathbf{U}_i\mathbf{U}_i'\right\} - \mathrm{vec}(\mathbf{UCOV})\mathrm{vec}(\mathbf{UCOV})'\right].$$

In the elliptic case $n\widehat{\mathbf{COV}}(\mathrm{vec}(\mathbf{UCOV})) \to_p \tau_1 \mathbf{C}_{p,p}$, and the statistic (which is valid in the wider model)

$$(\mathbf{C}_{p,p}\mathrm{vec}(\mathbf{UCOV}))'\widehat{\mathbf{COV}}(\mathrm{vec}(\mathbf{UCOV}))^{-}(\mathbf{C}_{p,p}\mathrm{vec}(\mathbf{UCOV}))$$

is asymptotically equivalent to $(n/\tau)Q^2$.

Next we introduce the shape matrix estimate $\mathbf{S} = \mathbf{S}(\mathbf{Y})$ corresponding to the spatial sign score and give its limiting distribution in the elliptic case. This estimate was already used to standardize the observations in the one sample location testing problem.

Definition 9.2. The Tyler shape estimate **S** based on spatial signs is the matrix that solves

$$\mathbf{UCOV}(\mathbf{Y}\mathbf{S}^{-1/2}) = \frac{1}{p}\mathbf{I}_p \tag{9.1}$$

The estimate was given in Tyler (1987) where the limiting distribution is also found.

Theorem 9.2. *Under elliptical symmetry with shape parameter Λ, the limiting distribution of the shape estimate* **S** *is given by*

$$\sqrt{n}\, vec(\mathbf{S} - \Lambda) \to_d N_{p^2}\left(0, (p+2)^2\tau\mathbf{C}_{p,p}(\Lambda)\right).$$

Here Λ and **S** *are normalized so that* $\det(\Lambda) = \det(\mathbf{S}) = 1$.

It is remarkable that in the elliptic case this estimate is distribution-free.

The case of unknown location μ is also considered in Tyler (1987). It is, for example, possible to replace μ with a $\sqrt{n}$-consistent estimate $\hat{\mu}$ without affecting the asymptotic properties of **UCOV** or **S**. As mentioned before, Hettmansperger and Randles (2002) propose a simultaneous estimation of the multivariate median μ and a shape matrix Λ.

9.5 Test and estimates based on TCOV

For the null hypothesis $H_0: \Lambda = \mathbf{I}_p$, the multivariate Kendall's tau-type rank test statistic $\mathbf{TCOV} = \mathbf{TCOV}(\mathbf{Y})$ is constructed in exactly the same way as the sign test statistic but for the pairwise differences. We denote the pairwise differences by $\mathbf{y}_{ij} = \mathbf{y}_i - \mathbf{y}_j$ and their spatial signs by $\mathbf{U}_{ij} = \mathbf{U}(\mathbf{y}_{ij})$, $1 \le i, j \le n$.

Definition 9.3. The Kendall's tau covariance matrix is defined as

$$\mathbf{TCOV} = \mathbf{AVE}_{i<j}\left\{\mathbf{U}_{ij}\mathbf{U}_{ij}'\right\}.$$

This matrix is introduced and studied in Visuri et al. (2000). Because vec($\mathbf{TCOV}$) is a U-statistic with bounded vector-valued kernel

$$\mathbf{h}(\mathbf{y}_i, \mathbf{y}_j) = \mathbf{U}_{ij} \otimes \mathbf{U}_{ij},$$

the limiting multinormality easily follows with

$$\mathbf{E}(\mathrm{vec}(\mathbf{TCOV})) = p^{-1}\mathrm{vec}(\mathbf{I}_p) \quad \text{and} \quad \mathbf{COV}(\mathrm{vec}(\mathbf{TCOV})) = \frac{\tau_F}{n}\mathbf{C}_{p,p} + o\left(\frac{1}{n}\right)$$

with some $\tau_F > 0$. In the general case, the covariance may be estimated by

$$\widehat{\mathbf{COV}}(\mathrm{vec}(\mathbf{TCOV})) = \frac{4}{n}\left[\mathbf{AVE}\left\{\mathbf{U}_{ij}\mathbf{U}_{ik}' \otimes \mathbf{U}_{ij}\mathbf{U}_{ik}'\right\} - \mathrm{vec}(\mathbf{TCOV})\mathrm{vec}(\mathbf{TCOV})'\right].$$

The test statistic based on Kendall's tau covariance matrix is defined as follows.

Definition 9.4. The *Kendall's tau test statistic* is defined as

$$Q^2 = Q^2(\mathbf{TCOV}) = |\mathbf{C}_{p,p}\mathbf{TCOV}|^2.$$

Again, it holds that $tr(\mathbf{TCOV}) = 1$ and so $Q^2 = 0$ only when $\mathbf{TCOV} = (1/p)\mathbf{I}_p$. Note that because the test statistics is based on pairwise differences, parameter μ need not be known.

Theorem 9.3. *Under the alternative sequence H_n, with a constant $\tau_F > 0$ depending on F,*

$$\frac{n}{\tau_F}Q^2 \to_d \chi^2_{(p+2)(p-1)/2}\left(\frac{1}{(p+2)^2\tau_F}Q^2(\mathbf{D})\right).$$

In practice, the distribution of the observations, or more specifically the distribution of the length of the observations, is not specified, and the coefficient τ_F is not known and has to be estimated. See Sirkiä et al. (2008). Alternatively, as $n\widehat{\mathbf{COV}}(\mathrm{vec}(\mathbf{TCOV})) \to_p \tau_F\mathbf{C}_{p,p}$ in the spherical case, one may use the statistic

$$(\mathbf{C}_{p,p}\mathrm{vec}(\mathbf{TCOV}))'\,\widehat{\mathbf{COV}}(\mathrm{vec}(\mathbf{TCOV}))^-\,(\mathbf{C}_{p,p}\mathrm{vec}(\mathbf{TCOV})).$$

The Kendall's tau test statistic naturally leads to a companion estimate of shape.

Definition 9.5. The *Dümbgen shape estimate* $\mathbf{S}$ based on the spatial signs of pairwise differences, is the one that solves

$$\mathbf{TCOV}(\mathbf{Y}\mathbf{S}^{-1/2}) = \frac{1}{p}\mathbf{I}_p.$$

The estimator was first introduced by Dümbgen (1998) and further studied by Sirkiä et al. (2007) as a member of a class of symmetrized M-estimates of scatter. The algorithm to calculate $\mathbf{S}$ is similar to that for calculating Tyler's estimate. The breakdown properties were considered in Dümbgen and Tyler (2005). The Dümbgen estimate is again affine equivariant in the sense that $\mathbf{S}(\mathbf{Y}\mathbf{A}') \propto \mathbf{A}\mathbf{S}(\mathbf{Y})\mathbf{A}'$.

The limiting distribution in the elliptic case is given in the following theorem.

Theorem 9.4. *At elliptical distribution with shape parameter* Λ*, the limiting distribution of the shape estimate* $\mathbf{S}$ *is given by*

$$\sqrt{n}vec(\mathbf{S}-\Lambda) \to_d N_{p^2}\left(0,(p+2)^2\tau_F\mathbf{C}_{p,p}(\Lambda)\right).$$

9.6 Tests and estimates based on RCOV

The Spearman's rho-type test statistic for the null hypothesis $H_0:\ \Lambda = \mathbf{I}_p$ is constructed in the same way as the spatial sign test statistic but using spatial rank covariance matrix $\mathbf{RCOV} = \mathbf{RCOV}(\mathbf{Y})$ instead of the spatial sign covariance matrix. Denote again $\mathbf{y}_{ij} = \mathbf{y}_i - \mathbf{y}_j$ and $\mathbf{U}_{ij} = \mathbf{S}(\mathbf{y}_{ij})$.

Definition 9.6. The spatial rank covariance matrix is defined as

$$\mathbf{RCOV} = \mathbf{AVE}\left\{\mathbf{R}_i\mathbf{R}_i'\right\} = \mathbf{AVE}\left\{\mathbf{U}_{ij}\mathbf{U}_{ik}'\right\}.$$

This matrix is considered in Marden (1999b) and Visuri et al. (2000). Now $\text{vec}(\mathbf{RCOV})$ is (up to a constant) asymptotically equivalent to a U-statistic with symmetric kernel

$$\mathbf{h}(\mathbf{y}_1,\mathbf{y}_2,\mathbf{y}_3) = \text{vec}\left(\mathbf{U}_{12}\mathbf{U}_{13}' + \mathbf{U}_{13}\mathbf{U}_{12}' + \mathbf{U}_{21}\mathbf{U}_{23}' + \mathbf{U}_{23}\mathbf{U}_{21}' + \mathbf{U}_{31}\mathbf{U}_{32}' + \mathbf{U}_{32}\mathbf{U}_{31}'\right),$$

covering all six possible permutations of the three arguments. Then under the null hypothesis,

$$\mathbf{E}(\mathbf{C}_{p,p}\text{vec}(\mathbf{RCOV})) = \mathbf{0} \quad \text{and} \quad \mathbf{COV}(\mathbf{C}_{p,p}\text{vec}(\mathbf{RCOV})) = \frac{\tau_F}{n}\mathbf{C}_{p.p} + o\left(\frac{1}{n}\right),$$

again with some constant τ_F. In the general case, the covariance may be estimated with

$$\widehat{\mathbf{COV}}(\mathbf{C}_{p,p}vec(\mathbf{RCOV})) =$$
$$\frac{9}{n}\left[\mathbf{AVE}\left\{\mathbf{h}(\mathbf{y}_i,\mathbf{y}_j,\mathbf{y}_k)\mathbf{h}(\mathbf{y}_i,\mathbf{y}_l,\mathbf{y}_m)'\right\} - \text{vec}(\mathbf{RCOV})\text{vec}(\mathbf{RCOV})'\right].$$

Definition 9.7. The *Spearman's rho test statistic* is defined as $Q^2 = Q^2(\mathbf{RCOV})$.

Note that the trace of **RCOV** is not fixed any more but varies from sample to sample. Moreover, even though the expected value of **RCOV** under the null hypothesis is always proportional to the identity matrix, the expected value of the trace depends on the distribution. Still, Q^2 is equal to zero only when **RCOV** is proportional to the identity matrix. The asymptotic distribution of Q^2 is given in the following.

Theorem 9.5. *Under the alternative sequence H_n,*

$$\frac{n}{\tau_F} Q^2 \to_d \chi^2_{(p+2)(p-1)/2}\left(\frac{(c_F^2)^2}{(p+2)^2\tau_F} Q^2(\mathbf{D})\right),$$

where τ_F and c_F^2 depend on the background distribution F.

Again, the coefficient τ_F needs to be estimated when the distribution of the data is not known. Alternatively, as $n\widehat{\mathbf{COV}}(\mathbf{C}_{p,p}\text{vec}(\mathbf{RCOV})) \to_p \tau_F \mathbf{C}_{p,p}$ in the spherical case, one may use the statistic

$$(\mathbf{C}_{p,p}\text{vec}(\mathbf{RCOV}))'\widehat{\mathbf{COV}}(\mathbf{C}_{p,p}\text{vec}(\mathbf{RCOV}))^{-}(\mathbf{C}_{p,p}\text{vec}(\mathbf{RCOV})),$$

which is asymptotically equivalent to $(n/\tau_F)Q^2$.

As in the previous sections, it is possible to define a shape estimate corresponding to the above test.

Definition 9.8. The shape estimate $\mathbf{S} = \mathbf{S}(\mathbf{Y})$ based on the rank covariance matrix is the one for which $Q^2(\mathbf{RCOV}(\mathbf{Y}\mathbf{S}^{-1/2})) = 0$, that is, for which

$$\mathbf{RCOV}(\mathbf{Y}\mathbf{S}^{-1/2}) \propto \mathbf{I}_p.$$

Unfortunately, there is no proof of the uniqueness or even existence of the solution. It seems to us, however, that one can use an iterative algorithm with steps

$$\mathbf{S} \leftarrow \mathbf{S}^{1/2}\mathbf{RCOV}\mathbf{S}^{1/2} \quad \text{and} \quad \mathbf{S} \leftarrow \frac{p}{tr(\mathbf{S})}\mathbf{S}.$$

In practice the algorithm always seems to yield a unique solution. The following theorem gives the limiting distribution of the shape estimator assuming that it is unique and $\sqrt{n}$-consistent.

Theorem 9.6. *Under the assumptions above, for an elliptical distribution with shape parameter Λ, the limiting distribution of the shape estimate $\mathbf{S}$ is given by*

$$\sqrt{n}\, vec(\mathbf{S} - \Lambda) \to_d N_{p^2}\left(0, (p+2)^2\tau_F(c_F^2)^{-1}\mathbf{C}_{p,p}(\Lambda)\right).$$

9.7 Limiting efficiencies

We next compare the sphericity tests based on **UCOV**, **TCOV**, and **RCOV** to the classical (modified) John's test. The modified John's test is based on the test statistic

$$Q^2_{J'} = \frac{np^2}{2(1+\kappa_F)}\, Q^2\left(\frac{\mathbf{COV}}{tr(\mathbf{COV})}\right),$$

where κ_F is the classical kurtosis measure of the marginal distribution. The limiting distribution of the modified John's test under the alternative sequence is derived in Hallin and Paindaveine (2006) and is given in the following.

Theorem 9.7. *Under the alternative sequence H_n,*

$$Q^2_{J'} \to_d \chi^2_{(p+2)(p-1)/2}\left(\frac{1}{2(1+\kappa_F)} Q^2(\mathbf{D})\right).$$

The limiting distributions of different test statistics are of the same type, therefore the efficiency comparisons may simply be based on their noncentrality parameters. The Pitman asymptotic relative efficiencies of tests based on **UCOV**, **TCOV** and **RCOV** with respect to the modified John's test (based on **COV**) reduce to

$$\frac{p(1+\kappa_F)}{p+2},\quad \frac{2(1+\kappa_F)}{(p+2)^2\tau_F},\quad \text{and}\quad \frac{2(c_F^2)^2(1+\kappa_F)}{(p+2)^2\tau_F}.$$

Recall that κ_F, τ_F, and c_F^2 are constants depending on the underlying distribution. Note that the Pitman AREs give the asymptotical relative efficiencies of the corresponding shape estimates as well.

In Table 9.1, the limiting efficiencies are given under t-distributions with some selected dimensions p and some degrees of freedom ν, with $\nu = \infty$ referring again to the multivariate normal case. Note that $\kappa_F = 0$ for the multivariate normal distribution, and $\kappa_F = 2/(\nu-4)$ for the multivariate t_ν distribution. Formulas for calculating the c_F^2 coefficients can be found in Möttönen et al. (1997). One can see that the

Table 9.1 Asymptotic relative efficiencies of tests (estimates) based on **UCOV**, **TCOV**, and **RCOV** relative to the test based on **COV** for different t-distribution cases with selected values of dimension p and degrees of freedom ν

	$p=2$			$p=4$			$p=5$		
ν	UCOV	TCOV	RCOV	UCOV	TCOV	RCOV	UCOV	TCOV	RCOV
5	1.50	2.43	2.42	2.00	2.62	2.56	2.14	2.71	2.63
8	0.75	1.26	1.25	1.00	1.32	1.30	1.07	1.35	1.31
15	0.59	1.04	1.04	0.79	1.07	1.06	0.84	1.08	1.07
∞	0.50	0.93	0.95	0.67	0.95	0.97	0.71	0.95	0.99

rank-based tests based on **TCOV** and **RCOV** behave very similarly and are highly efficient even in the normal case. The test based on **UCOV** is less efficient than the rank-based tests but still outperforms the classical test for heavy-tailed distributions. Note that the efficiencies increase with dimension. See Sirkiä et al. (2008) for a more complete discussion and for the finite-sample efficiencies.

9.8 Examples

Example 9.1. **Cork boring data: The tests and estimates for shape.** To illustrate the estimated shape matrices we plot the corresponding estimates of the 50 % tolerance regions. The estimated tolerance regions based on a location estimate **T** and a shape estimate **S** are constructed as follows. First calculate the squared Mahalanobis distances based on **S** and **T**; that is,

$$r_i^2 = |\mathbf{y}_i - \mathbf{M}|_{\mathbf{S}}^2 = (\mathbf{y}_i - \mathbf{T})'\mathbf{S}^{-1}(\mathbf{y}_i - \mathbf{T}), \quad i = 1,...,n.$$

The estimated 50 % tolerance region is then the ellipsoid

$$\{\mathbf{y} : |\mathbf{y} - \mathbf{M}|_{\mathbf{S}}^2 \leq Med\{r_1^2,...,r_n^2\}\}.$$

We compare the tolerance ellipsoids for the shape matrices based on **COV**, **UCOV** (Tyler's shape), and **TCOV** (Dümbgen's shape). The corresponding location estimates are the sample mean, the (affine equivariant) spatial median, and the (affine equivariant) Hodges-Lehman estimate. Note that if we are interested in the shape matrix, only the shape (not the size or location) of the tolerance ellipsoid is relevant. The shape should be circular or spherical in the case that $\Lambda = \mathbf{I}_p$. The tolerance ellipsoids for the 3-variate cork boring data are given in Figure 9.1.

The figures are obtained using the following R code.

```
> data(cork)
> cork_3v <- sweep(cork[,2:4], 1, cork[,1], "-")
> colnames(cork_3v) <- c("E_N", "S_N", "W_N")
>
> EST1 <- list(location = colMeans(cork_3v),
  scatter = cov(cork_3v),
  est.name = "COV")
>
> HR.cork_3v <- HR.Mest(cork_3v)
> EST2 <- list(location = HR.cork_3v$center,
  scatter = HR.cork_3v$scatter,
  est.name = "Tyler")
>
```

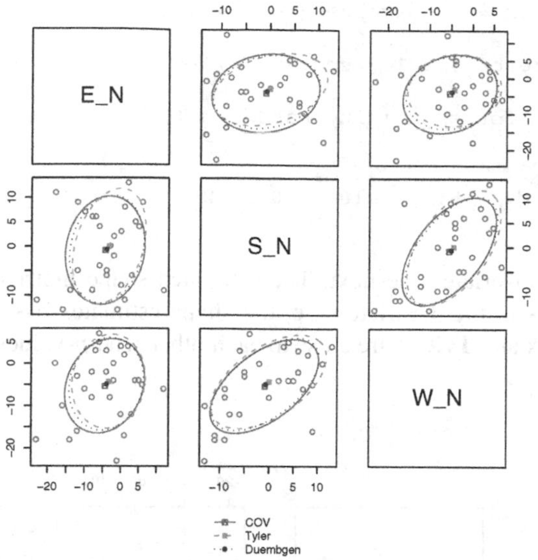

Fig. 9.1 The shape matrix estimates for the 3-variate data.

```
> EST3 <- list(location = mv.1sample.est(cork_3v, score="rank",
  stand = "inner")$location, scatter = duembgen.shape(cork_3v),
  est.name = "Duembgen")
>
> plotShape(EST1, EST2, EST3, cork_3v, lty.ell = 1:3,
  pch.ell = 14:16, level = 0.5)
```

The ellipsoids do not seem spherical but only the test based on the regular covariance matrix gets a small p-value. (The sample size is too small for a reliable inference on the shape parameter.) The p-values are as follows.

```
> mv.shape.test(cork_3v)

        Mauchly test for sphericity

data:  cork_3v
L = 0.0037, df = 5, p-value = 0.04722

> mv.shape.test(cork_3v, score= "si")

        Test for sphericity based on UCOV

data:  cork_3v
```

```
Q2 = 6.023, df = 5, p-value = 0.3040

> mv.shape.test(cork_3v, score= "sy")

        Test for sphericity based on TCOV

data:  cork_3v
Q2 = 9.232, df = 5, p-value = 0.1002
```

Consider the bivariate case next. The estimated shape matrices are illustrated in Figure 9.2. As in the 3-variate case, the shape estimates based on the regular covariance matrix and **TCOV** are close to each other. The *p*-values are now

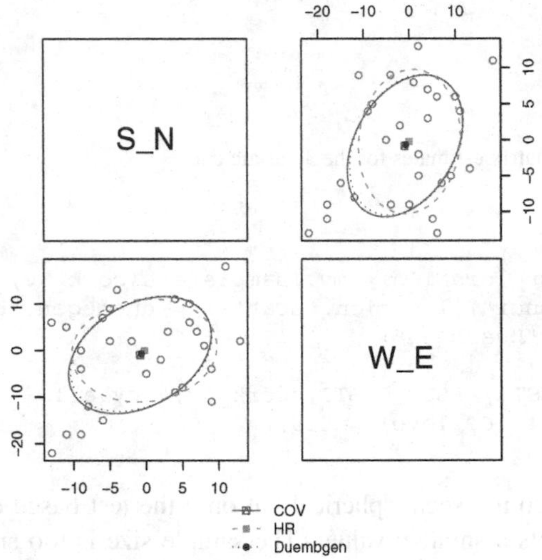

Fig. 9.2 The shape matrix estimates for the 2-variate data.

```
> mv.shape.test(cork_2v)

        Mauchly test for sphericity

data:  cork_2v
L = 0.0748, df = 2, p-value = 0.07478
```

```
> mv.shape.test(cork_2v, score= "si")

        Test for sphericity based on UCOV

data:  cork_2v
Q2 = 0.1885, df = 2, p-value = 0.91

> mv.shape.test(cork_2v, score= "sy")

        Test for sphericity based on TCOV

data:  cork_2v
Q2 = 2.637, df = 2, p-value = 0.2675

>
```

Example 9.2. **Comparison of the tests and estimates in the t_3 case.** To illustrate the finite sample efficiencies of the estimates for a heavy-tailed distribution we generated a random sample of size $n = 150$ from a spherical 3-variate t_3 distribution. The null hypothesis $H_0 : \Lambda = \mathbf{I}_p$ is thus true. The three shape estimates based on **COV**, **UCOV**, and **TCOV** are illustrated in Figure 9.3. The R code for getting the figure follows.

```
> set.seed(1234)
> X<-rmvt(150, diag(3),3)
>
>
> EST1 <- list(location = colMeans(X), scatter = cov(X),
  est.name = "COV")
>
> HR.X<-HR.Mest(X)
> EST2 <- list(location = HR.X$center, scatter = HR.X$scatter,
  est.name = "Tyler")
>
> EST3 <- list(location = mv.1sample.est(X, score = "rank",
  stand = "inner")$location,
  scatter = duembgen.shape(X), est.name = "Duembgen")
>
> plotShape(EST1, EST2, EST3, X, lty.ell = 1:3,
  pch.ell = 14:16, level = 0.95)
```

The shape of the regular covariance matrix clearly differs most from the spherical shape. This can also be seen from the p-values below. The regular covariance matrix does not seem too reliable in the heavy-tailed distribution case.

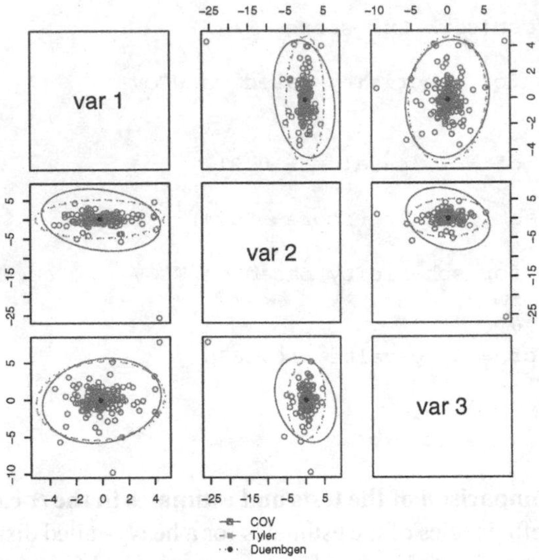

Fig. 9.3 Shape estimates for a sample from a 3-variate t_3 distribution.

```
> mv.shape.test(X)

        Mauchly test for sphericity

data:  X
L = 0, df = 5, p-value = 2.371e-13

> mv.shape.test(X, score= "si")

        Test for sphericity based on UCOV

data:  X
Q2 = 2.476, df = 5, p-value = 0.7801

> mv.shape.test(X, score= "sy")

        Test for sphericity based on TCOV

data:  X
Q2 = 1.323, df = 5, p-value = 0.9326
```

9.9 Principal component analysis based on spatial signs and ranks

Let again $\mathbf{Y} = (\mathbf{y}_1, ..., \mathbf{y}_n)'$ be a random sample from a p-variate elliptical distribution with the cumulative distribution function F. Write

$$\mathbf{COV}(F) = \mathbf{ODO}'$$

for the eigenvector and eigenvalue decomposition of the covariance matrix. Thus $\mathbf{O}$ is the matrix of eigenvectors and $\mathbf{D}$ is the diagonal matrix of eigenvalues of $\mathbf{COV}(F)$. The orthogonal matrix $\mathbf{O}$ is then used to transform the random vector $\mathbf{y}_i$ to a new coordinate system

$$\mathbf{z}_i = \mathbf{O}'\mathbf{y}_i, \quad i = 1, ..., n.$$

The components of $\mathbf{z}_i$ in this new coordinate system are called the principal components. The principal components are then uncorrelated and ordered according to their variances (diagonal elements of $\mathbf{D}$). In the multivariate normal case the principal components are independent. In principal component analysis (PCA) one wishes to estimate the transformation $\mathbf{O}$ to principal components as well as the variances $\mathbf{D}$. PCA is often used to reduce the dimension of the original vector from $p = p_1 + p_2$ to p_1, say. If

$$\mathbf{O} = (\mathbf{O}_1, \mathbf{O}_2),$$

where $\mathbf{O}_1$ is a $p \times p_1$ matrix and $\mathbf{O}_2$ a $p \times p_2$ matrix, then the original observations

$$\mathbf{y}_i = \mathbf{O}_1\mathbf{O}_1'\mathbf{y}_i + \mathbf{O}_2\mathbf{O}_2'\mathbf{y}_i, \quad i = 1, ..., n,$$

are approximated by the first part

$$\hat{\mathbf{y}}_i = \mathbf{O}_1\mathbf{O}_1'\mathbf{y}_i = \mathbf{o}_1 z_{i1} + \cdots + \mathbf{o}_{p_1} z_{ip_1}, \quad i = 1, ..., n.$$

As shown before in Theorem 4.4, the eigenvalue and eigenvector decompositions of

UCOV and **TCOV** satisfy, for the elliptical distribution F,

$$\mathbf{UCOV}(F) = \mathbf{OD}_U\mathbf{O}' \quad \text{and} \quad \mathbf{TCOV}(F) = \mathbf{OD}_T\mathbf{O}'$$

with the same principal component transformation. The same is naturally true for the scatter or shape matrix functionals $\mathbf{S}$ that are based on **UCOV** and **TCOV**, namely for Tyler's shape estimate and Dümbgen shape estimate. This means that the eigenvectors of the sample matrices can be used to estimate the unknown population eigenvectors. Locantore et al. (1999) and Marden (1999b) proposed the use of the spatial signs and ranks for a simple robust alternative of classical PCA. See also Visuri et al. (2000) and Croux et al. (2002).

All the scatter and shape matrices mentioned above are root-n consistent and have a limiting multivariate normal distribution. We next find the limiting distribution of the corresponding sample eigenvalue matrix, that is, an estimate of the principal component transformation. For simplicity we assume that $\mathbf{O} = \mathbf{I}_p$ and that the eigenvalues listed in $\mathbf{D}$ are distinct. (The limiting distribution in the general case is found simply by using the rotation equivariance properties of the estimates.) Then we have the following.

Theorem 9.8. *Let* $\mathbf{S}$ *be a positive definite symmetric* $p \times p$ *random matrix such that the limiting distribution of* $\sqrt{n}(\mathbf{S} - \mathbf{D})$ *is a* p^2*-variate (singular) normal distribution with zero mean vector. Let* $\mathbf{S} = \hat{\mathbf{O}}\hat{\mathbf{D}}\hat{\mathbf{O}}'$ *be the eigenvalue and eigenvector decomposition of S. Then the limiting distributions of*

$$\sqrt{n}\, vec(\hat{\mathbf{O}} - \mathbf{I}_p) \quad and \quad \sqrt{n}\, vec(\hat{\mathbf{D}} - \mathbf{D})$$

are both multivariate normal and given by

$$\sqrt{n}\, vec(\mathbf{S} - \mathbf{D}) = ((\mathbf{D} \otimes \mathbf{I}_p) - (\mathbf{I}_p \otimes \mathbf{D}))\sqrt{n}\, vec(\hat{\mathbf{O}} - \mathbf{I}_p) + \sqrt{n}\, vec(\hat{\mathbf{D}} - \mathbf{D}) + o_P(1).$$

If we are interested in the limiting distribution of $\hat{\mathbf{O}}_{ij}$, we then obtain

$$\sqrt{n}\hat{\mathbf{S}}_{ij} = (D_{ii} - D_{jj})\sqrt{n}\hat{\mathbf{O}}_{ij} + o_P(1), \quad i \neq j.$$

and

$$\sqrt{n}(\hat{\mathbf{O}}_{ii} - 1) = o_P(1), \quad i = 1, ..., n.$$

The efficiencies of the shape matrices then give the efficiencies for the eigenvectors as well.

Example 9.3. **Principal component analysis on the air pollution dataset.** We apply the robust principal component analysis based on Tyler's and Dümbgen's shape matrices for the air pollution data in the United States. In the dataset we use 6 measurements on 41 cities in the United States. The data were originally collected to build a regression model for pollution measurements (SO2) using these 6 explaining variables. Two of the 6 explaining variables were related to human population (LargeF, Pop); four were climate variables (NegTemp, Wind, AvRain, DaysRain). See Section 3.3 in Everitt (2005) for more details and the use of PCA for this dataset. For the principal component analysis, we first rescaled all the marginal variables with the median absolute deviation (MAD). The results are then independent on the measurement units used for the original variables. (The PCA is not affine invariant, however.) See Figure 9.4 for the transformed data set.

The R code for getting the figure follows.

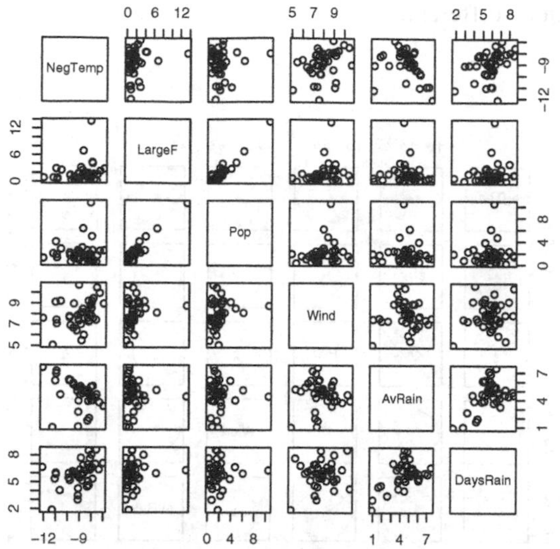

Fig. 9.4 Air pollution dataset. The marginal variables are standardized with MAD.

```
> data(usair)
> pairs(usair)
>
> # the dataset as used in Everitt + giving the variables names
>
> usair2 <- usair[,-1]
> usair2 <- transform(usair2, x1 = -x1)
> colnames(usair2) <- c("NegTemp", "LargeF", "Pop", "Wind",
  "AvRain", "DaysRain")
>
> mads <- apply(usair2, 2, mad)
> usair3 <- sweep(usair2,2,mads, "/")
>
> pairs(usair3)
>
```

The next step is to calculate the three shape matrices: Tyler's and Dümbgen's shape matrices and the one based on the regular covariance matrix. (The corresponding transformations standardize **UCOV**, **TCOV** and **COV**, resp.) As some of the marginal distributions are strongly skewed, the shape of the regular covariance matrix clearly differs most from the spherical shape. Unlike the spatial sign- and rank-based shape matrices, the covariance matrix is very sensitive to heavy tails, which can be clearly seen from Figure 9.5. One can then expect that the results in the PCA can be quite different.

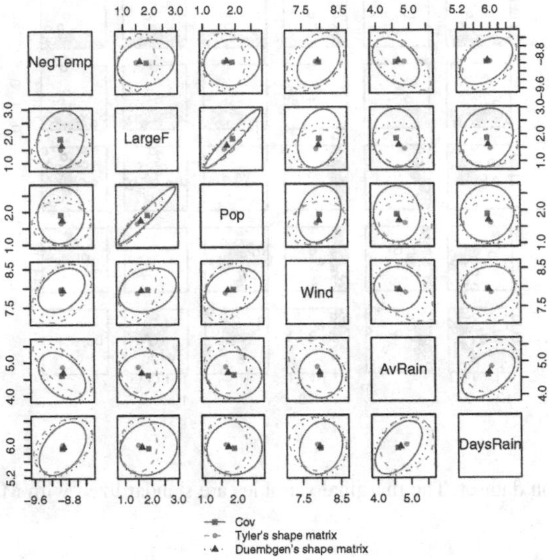

Fig. 9.5 Air pollution dataset: Shape estimates.

```
>
> COV <- cov(usair3)
> SI <- HR.Mest(usair3)
> Dumb <-duembgen.shape(usair3)
> rank.inner <-rank.shape(usair3)
>
> aff.HL <- mv.1sample.est(usair3, score = "rank",
  stand = "inner")$location
>
> classical <- list(location= colMeans(usair3),
  scatter=  COV / sum(diag(COV)) * 6, est.name="Cov")
> signs.inner <- list(location= SI$center,
  scatter=  SI$scatter / sum(diag(SI$scatter)) * 6,
  est.name="Tyler's shape matrix")
> symm.signs.inner <- list(location=aff.HL,
  scatter=  Dumb / sum(diag(Dumb)) * 6 ,
  est.name="Duembgen's shape matrix")
> ranks.inner <- list(location= aff.HL, scatter=
```

```
  rank.inner / sum(diag(rank.inner)) * 6,
  est.name="Rank shape matrix")
>
>
> plotShape(classical, signs.inner, symm.signs.inner,
  lty.ell= 1:3, pch.ell=15:17,
  + x.legend= -3 ,y.legend= -1.2,
  labels=colnames(usair3), cex.labels = 1.5)
```

In the following the R-function with the three scores is applied to get the results in the corresponding PCA. We first compare the similarity of the results by calculating the correlations between the principal components coming from different approaches. The correlations show some similarity between the rank- and sign-based solutions whereas the regular PCA solution differs from the others.

```
>
> PCA.identity <- mvPCA(usair3, score = "identity")
> PCA.signs.inner <- mvPCA(usair3, score = "sign",
  estimate= "inner")
> PCA.symm.signs.inner <- mvPCA(usair3, score = "sym",
  estimate= "inner")
> round(cor(PCA.signs.inner$scores,PCA.identity$scores),2)
       Comp.1 Comp.2 Comp.3 Comp.4 Comp.5 Comp.6
Comp.1   0.81  -0.34  -0.47   0.09   0.01   0.00
Comp.2  -0.91  -0.16  -0.37  -0.07  -0.01   0.01
Comp.3  -0.41  -0.71   0.36   0.45  -0.04   0.00
Comp.4   0.22  -0.75   0.30  -0.54  -0.01  -0.01
Comp.5  -0.58  -0.22   0.14   0.08   0.74  -0.20
Comp.6   0.14  -0.14   0.12  -0.03   0.38   0.89
> round(cor(PCA.symm.signs.inner$scores,PCA.identity$scores),2)
       Comp.1 Comp.2 Comp.3 Comp.4 Comp.5 Comp.6
Comp.1   0.99  -0.10  -0.12   0.04   0.01   0.00
Comp.2  -0.45  -0.68  -0.57   0.03   0.00   0.00
Comp.3   0.06  -0.73   0.67   0.16  -0.01   0.00
Comp.4  -0.23   0.34  -0.16   0.90  -0.01   0.00
Comp.5  -0.32  -0.11   0.07   0.04   0.93  -0.14
Comp.6   0.17  -0.04   0.05   0.00   0.23   0.96
> round(cor(PCA.symm.signs.inner$scores,
            PCA.signs.inner$scores),2)
       Comp.1 Comp.2 Comp.3 Comp.4 Comp.5 Comp.6
Comp.1   0.89  -0.84  -0.36   0.24  -0.55   0.14
Comp.2   0.14   0.73   0.48   0.23   0.33  -0.04
Comp.3   0.00  -0.19   0.80   0.67   0.22   0.18
Comp.4  -0.14   0.16   0.21  -0.84   0.11  -0.12
Comp.5  -0.24   0.27   0.22   0.00   0.94   0.21
Comp.6   0.12  -0.17  -0.03   0.07  -0.10   0.98
```

We then compare the results based on the regular shape matrix with those based on Tyler's shape matrix. The principal components coming from the regular shape matrix may now be easier to interpret. The first one is related to the human population, the second one to the rain conditions, the third one to other climate variables, and the fourth one to the wind. The robust shape matrices cut down the effects of few outlying observations, the few cities with very high population and manufacturing as well as the few cities with a hot climate and low precipitation. The results coming from the robust PCA as reported below (based on Tyler's shape matrix) therefore differ a lot from the results of the regular PCA. The first principal component may be related to the climate in general, the second one to the human population, the third one to the rain, and the fourth one to the wind.

```
> summary(PCA.identity, loadings=TRUE)
Importance of components:
                         Comp.1 Comp.2 Comp.3  Comp.4
Proportion of Variation  0.5527 0.1991 0.1506 0.07555
Cumulative Proportion    0.5527 0.7518 0.9024 0.97792

                          Comp.5   Comp.6
                         0.01368 0.008397
                         0.99160 1.000000

Loadings:
         Comp.1 Comp.2 Comp.3 Comp.4 Comp.5 Comp.6
NegTemp                 0.703  0.227  0.556  0.371
LargeF   -0.780                       0.272 -0.554
Pop      -0.604        -0.176        -0.330  0.702
Wind     -0.121         0.405 -0.893 -0.115
AvRain          -0.755 -0.383 -0.188  0.461  0.188
DaysRain        -0.649  0.402  0.327 -0.531 -0.152
> PCA.signs.inner
PCA for usair3 based on Tyler's shape matrix

Standardized eigenvalues:
Comp.1 Comp.2 Comp.3 Comp.4 Comp.5 Comp.6
2.4436 1.5546 1.1125 0.6872 0.1193 0.0829

 6 variables and 41 observations.
> summary(PCA.signs.inner, loadings=TRUE)
Importance of components:
                         Comp.1 Comp.2 Comp.3 Comp.4
Proportion of Variation  0.4073 0.2591 0.1854 0.1145
Cumulative Proportion    0.4073 0.6664 0.8518 0.9663

                         Comp.5  Comp.6
                         0.01988 0.01382
                         0.98618 1.00000

Loadings:
         Comp.1 Comp.2 Comp.3 Comp.4 Comp.5 Comp.6
NegTemp  -0.453 -0.402  0.353         0.458  0.546
LargeF   -0.389  0.564  0.164 -0.164  0.520 -0.454
```

```
Pop        -0.302  0.602          -0.146 -0.476  0.541
Wind       -0.481         -0.342   0.798
AvRain      0.566  0.392   0.153   0.480  0.397  0.338
DaysRain                   0.838   0.289 -0.358 -0.286
> plot(PCA.signs.inner)
```

9.10 Other approaches

Hallin and Paindaveine (2006) proposed test statistics for sphericity based on spatial sign vectors $\mathbf{u}_i = |\mathbf{y}_i|^{-1}\mathbf{y}_i$ and the ranks of lengths $r_i = |\mathbf{y}_i|$, $i = 1,...,n$. Their test statistics are of the same form as that defined in John (1971), but instead of the sample covariance matrix, they use $\mathbf{AVE}\{K(R_i/(n+1))\mathbf{u}_i\mathbf{u}_i'\}$, where $K = K_g$ is a score function corresponding to spherical density g and R_i denotes the rank of r_i among $r_1,...,r_n$, $i = 1,...,n$. The Hallin and Paindaveine (2006) tests appear to be valid without any moments assumptions and asymptotically optimal if $f = g$.

Chapter 10
Multivariate tests of independence

Abstract Multivariate extensions of the quadrant test by Blomqvist (1950) and Kendall's tau and Spearman's rho statistics are discussed. Asymptotic theory is given to approximate the null distributions as well as to calculate limiting Pitman efficiencies. The tests are compared to the classical Wilks' (Wilks, 1935) test.

10.1 The problem and a general strategy

In the analysis of multivariate data, it is often of interest to find potential relationships among subsets of the variables. In the air pollution data example 9.3, for instance, two variables are related to human population and four variables represent climate attributes. It may be of interest to find out whether the two sets of variables are related. This requires a test of independence between pairs of vectors. The vectors may then have different measurement scales and dimensions.

Let $(\mathbf{X},\mathbf{Y})$ be a random sample of size n from a p-variable distribution where $\mathbf{X}$ is an $n\times r$ matrix of measurements of r variables and $\mathbf{Y}$ is an $n\times s$ matrix of measurements of remaining s variables, $r+s=p$. As before, we write

$$\mathbf{X}=(\mathbf{x}_1,...,\mathbf{x}_n)' \quad \text{and} \quad \mathbf{Y}=(\mathbf{y}_1,...,\mathbf{y}_n)'.$$

As $(\mathbf{X},\mathbf{Y})$ is a random sample, the rows are independent. Our null hypothesis of the independence of the x- and y-variables can be written as

$$H_0: \ \mathbf{X} \text{ and } \mathbf{Y} \text{ are independent.}$$

We wish to use general score functions $\mathbf{T}_x(\mathbf{x})$ and $\mathbf{T}_y(\mathbf{y})$ in the test construction. We again use the identity score function, the spatial sign score function, and the spatial rank score functions so that $\mathbf{T}_x(\mathbf{x})$ is an r-vector and $\mathbf{T}_y(\mathbf{y})$ is an s-vector,

H. Oja, *Multivariate Nonparametric Methods with R: An Approach Based on Spatial Signs and Ranks*, Lecture Notes in Statistics 199, DOI 10.1007/978-1-4419-0468-3_10,

respectively. As explained in Section 4.1, the observations in $\mathbf{X}$ may be replaced (i) by their outer centered and standardized scores or (ii) by their inner centered and standardized scores, and similarly and independently with $\mathbf{Y}$. If we use inner centering and standardization, we transform

$$(\mathbf{X},\mathbf{Y}) \;\rightarrow\; (\hat{\mathbf{T}}_{\mathbf{X}},\hat{\mathbf{T}}_{\mathbf{Y}}).$$

where

$$\hat{\mathbf{T}}_{\mathbf{X}}'\mathbf{1}_n = \hat{\mathbf{T}}_{\mathbf{Y}}'\mathbf{1}_n = 0,$$

and

$$r\cdot\hat{\mathbf{T}}_{\mathbf{X}}'\hat{\mathbf{T}}_{\mathbf{X}} = |\hat{\mathbf{T}}_{\mathbf{X}}|^2\mathbf{I}_r \quad \text{and} \quad s\cdot\hat{\mathbf{T}}_{\mathbf{Y}}'\hat{\mathbf{T}}_{\mathbf{Y}} = |\hat{\mathbf{T}}_{\mathbf{Y}}|^2\mathbf{I}_s.$$

Here, as before,

$$|\hat{\mathbf{T}}_{\mathbf{X}}|^2 = tr(\hat{\mathbf{T}}_{\mathbf{X}}'\hat{\mathbf{T}}_{\mathbf{X}}) \quad \text{and} \quad |\hat{\mathbf{T}}_{\mathbf{Y}}|^2 = tr(\hat{\mathbf{T}}_{\mathbf{Y}}'\hat{\mathbf{T}}_{\mathbf{Y}}).$$

Under general assumptions (which should be checked separately for each score function) and under the null hypothesis, the limiting distribution of

$$Q^2 = Q^2(\mathbf{X},\mathbf{Y}) = nrs\frac{|\hat{\mathbf{T}}_{\mathbf{X}}'\hat{\mathbf{T}}_{\mathbf{Y}}|^2}{|\hat{\mathbf{T}}_{\mathbf{X}}|^2|\hat{\mathbf{T}}_{\mathbf{Y}}|^2}$$

is then a chi-square distribution with rs degrees of freedom. As the null hypothesis says that

$$(\mathbf{X},\mathbf{PY}) \sim (\mathbf{X},\mathbf{Y})$$

for all $n\times n$ permutation matrices $\mathbf{P}$, the p-value from a conditionally distribution-free permutation test is obtained as

$$\mathbf{E}_{\mathbf{P}}\left\{I\left(Q^2(\mathbf{X},\mathbf{PY}) \geq Q^2(\mathbf{X},\mathbf{Y})\right)\right\}.$$

Note also that, as the inner and outer centering and standardization are permutation invariant,

$$Q^2(\mathbf{X},\mathbf{PY}) = nrs\frac{|\hat{\mathbf{T}}_{\mathbf{X}}'\mathbf{P}\hat{\mathbf{T}}_{\mathbf{Y}}|^2}{|\hat{\mathbf{T}}_{\mathbf{X}}|^2|\hat{\mathbf{T}}_{\mathbf{Y}}|^2}$$

so that simply $|\hat{\mathbf{T}}_{\mathbf{X}}'\mathbf{P}\hat{\mathbf{T}}_{\mathbf{Y}}|^2$ can be used in practice to find the permutation p-value.

10.2 Wilks' and Pillai's tests

The classical parametric test due to Wilks (1935) is the likelihood ratio test statistic in the multivariate normal model and is based on

$$W = W(\mathbf{X},\mathbf{Y}) = \frac{\det(\mathbf{COV})}{\det(\mathbf{COV}_{11})\det(\mathbf{COV}_{22})}$$

using the partitioned covariance matrix

$$\mathbf{COV}(\mathbf{X},\mathbf{Y}) = \mathbf{COV} = \begin{pmatrix} \mathbf{COV}_{11} & \mathbf{COV}_{12} \\ \mathbf{COV}_{21} & \mathbf{COV}_{22} \end{pmatrix},$$

where $\mathbf{COV}(\mathbf{X}) = \mathbf{COV}_{11}$ is an $r \times r$ matrix, and $\mathbf{COV}(\mathbf{Y}) = \mathbf{COV}_{22}$ is an $s \times s$ matrix. The test is optimal under the multivariate normal model. Under H_0 with finite fourth moments the test statistic $-n \log W \to_d \chi^2_{rs}$.

Another classical test for independence is Pillai's trace test statistic (Pillai (1955)) which uses

$$W^* = W^*(\mathbf{X},\mathbf{Y}) = tr\left(\mathbf{COV}_{11}^{-1}\mathbf{COV}_{12}\mathbf{COV}_{22}^{-1}\mathbf{COV}_{21}\right).$$

Pillai's trace statistic and Wilks' test statistic are asymptotically equivalent in the sense that if the fourth moments exist and the null hypothesis is true then

$$nW^* - n\log W \;\to_P 0.$$

If we follow our general strategy with the identity score function ($\mathbf{T}_x(\mathbf{x}) = \mathbf{x}$ and $\mathbf{T}_y(\mathbf{y}) = \mathbf{y}$) then it is easy to see that

$$Q^2(\mathbf{X},\mathbf{Y}) = nW^*(\mathbf{X},\mathbf{Y});$$

that is, Q^2 is the Pillai trace statistic.

Muirhead (1982) examined the effect of the group of affine transformations on this problem. The Wilks and Pillai tests are invariant under the group of affine transformations

$$(\mathbf{X},\mathbf{Y}) \to (\mathbf{X}\mathbf{H}_x' + \mathbf{1}_n\mathbf{b}_x', \mathbf{Y}\mathbf{H}_y' + \mathbf{1}_n\mathbf{b}_y')$$

for arbitrary r and s vectors $\mathbf{b}_x$ and $\mathbf{b}_y$ and for arbitrary nonsingular $r \times r$ and $s \times s$ matrices $\mathbf{H}_x$ and $\mathbf{H}_y$, respectively. Thus the performance of the tests does not depend on the variance-covariance structures of either $\mathbf{X}$ ($\mathbf{COV}_{11}$) or $\mathbf{Y}$ ($\mathbf{COV}_{22}$). This characteristic generally improves the power and control of α-levels.

10.3 Tests based on spatial signs and ranks

In the following we describe the tests that generalize the popular univariate tests due to Blomqvist (1950), Spearman (1904), and Kendall (1938) to any dimensions r and s. The tests provide practical and robust but still efficient alternatives to multivariate normal theory methods.

Extension of Blomqvist quadrant test. The test is thus based on the r- and s-variate spatial sign scores $\mathbf{U}(\mathbf{x})$ and $\mathbf{U}(\mathbf{y})$. To make the test statistic affine invariant, we first construct inner centered and standardized spatial signs separately for **X** and **Y** and transform

$$(\mathbf{X},\mathbf{Y}) \to (\hat{\mathbf{U}}_\mathbf{X},\hat{\mathbf{U}}_\mathbf{Y}).$$

(The outer centering and standardization yield procedures that are only rotation invariant.) Recall that the inner centering and standardization are accomplished by the simultaneous Hettmansperger-Randles estimates of multivariate location and shape; see Definition 6.4. For inner standardized spatial signs satisfy

$$\hat{\mathbf{U}}_\mathbf{X}'\mathbf{1}_n = \mathbf{0} \quad \text{and} \quad r\cdot\hat{\mathbf{U}}_\mathbf{X}'\hat{\mathbf{U}}_\mathbf{X} = n\mathbf{I}_r$$

and similarly

$$\hat{\mathbf{U}}_\mathbf{Y}'\mathbf{1}_n = \mathbf{0} \quad \text{and} \quad s\cdot\hat{\mathbf{U}}_\mathbf{Y}'\hat{\mathbf{U}}_\mathbf{Y} = n\mathbf{I}_s.$$

The test statistic is then

$$Q^2 = Q^2(\mathbf{X},\mathbf{Y}) = \frac{rs}{n}|\hat{\mathbf{U}}_\mathbf{X}'\hat{\mathbf{U}}_\mathbf{Y}|^2.$$

Consider next the limiting null distribution of Q^2. We assume that

$$(\mathbf{X},\mathbf{Y}) = (\mathbf{1}_n\mu_x' + \varepsilon_x\Omega_x', \mathbf{1}_n\mu_y' + \varepsilon_y\Omega_y'),$$

where $(\varepsilon_x,\varepsilon_y)$ is a random sample from a standardized distribution such that

$$\mathbf{E}\{\mathbf{U}(\varepsilon_{xi})\} = \mathbf{0} \quad \text{and} \quad r\cdot\mathbf{E}\{\mathbf{U}(\varepsilon_{xi})\mathbf{U}(\varepsilon_{xi})'\} = \mathbf{I}_r$$

and

$$\mathbf{E}\{\mathbf{U}(\varepsilon_{yi})\} = \mathbf{0} \quad \text{and} \quad s\cdot\mathbf{E}\{\mathbf{U}(\varepsilon_{yi})\mathbf{U}(\varepsilon_{yi})'\} = \mathbf{I}_s.$$

Note that, for $(\varepsilon_x)_i$ and $(\varepsilon_y)_i$ separately, this is a wider model than the model (B2) discussed in Chapter 2. The spatial median and Tyler's shape matrix of ε_{xi} (and similarly of ε_{yi}) is a zero vector and an identity matrix, respectively. If $\hat{\mu}_x$ and $\hat{\mu}_y$ are any root-n consistent estimates of μ_x and μ_y, and $\hat{\Sigma}_x$ and $\hat{\Sigma}_y$ are any root-n consistent estimates of $\Sigma_x = \Omega_x\Omega_x'$ and $\Sigma_y = \Omega_y\Omega_y'$ (up to a constant), respectively, and

$$(\hat{\mathbf{U}}_\mathbf{X})_i = \mathbf{U}\left(\hat{\Sigma}_x^{-1/2}(\mathbf{x}_i - \hat{\mu}_x)\right) \quad \text{and} \quad (\hat{\mathbf{U}}_\mathbf{Y})_i = \mathbf{U}\left(\hat{\Sigma}_y^{-1/2}(\mathbf{y}_i - \hat{\mu}_y)\right),$$

$i = 1,\ldots,n$, one can show as in Taskinen et al. (2003) using the expansions in Section 6.1.1 that, under the null hypothesis, $Q^2 \to_d \chi^2_{rs}$. (It naturally remains to show that the Hettmansperger-Randles estimate used in our approach is root-n consistent.)

Extension of Spearman's rho. The test is based on the spatial rank scores $\mathbf{R}_\mathbf{X}(\mathbf{x})$ and $\mathbf{R}_\mathbf{Y}(\mathbf{y})$ with dimensions r and s, respectively. We then calculate the inner standardized spatial ranks separately for **X** and **Y** and transform

$$(\mathbf{X},\mathbf{Y}) \to (\hat{\mathbf{R}}_\mathbf{X},\hat{\mathbf{R}}_\mathbf{Y}).$$

Note that, in this case, the scores are automatically centered, and the inner centering is therefore not needed. For inner standardized spatial ranks

$$\hat{\mathbf{R}}_{\mathbf{X}}'\mathbf{1}_n = \mathbf{0} \quad \text{and} \quad r\cdot \hat{\mathbf{R}}_{\mathbf{X}}'\hat{\mathbf{R}}_{\mathbf{X}} = |\hat{\mathbf{R}}_{\mathbf{X}}|^2\mathbf{I}_r$$

and

$$\hat{\mathbf{R}}_{\mathbf{Y}}'\mathbf{1}_n = \mathbf{0} \quad \text{and} \quad s\cdot \hat{\mathbf{R}}_{\mathbf{Y}}'\hat{\mathbf{R}}_{\mathbf{Y}} = |\hat{\mathbf{R}}_{\mathbf{Y}}|^2\mathbf{I}_s.$$

The test statistic is then

$$Q^2 = Q^2(\mathbf{X},\mathbf{Y}) = nrs\frac{|\hat{\mathbf{R}}_{\mathbf{X}}'\hat{\mathbf{R}}_{\mathbf{Y}}|^2}{|\hat{\mathbf{R}}_{\mathbf{X}}|^2|\hat{\mathbf{R}}_{\mathbf{Y}}|^2}.$$

Next consider the limiting null distribution of Q^2. For considering the asymptotic behavior of the test, we assume that there are (up to a constant) unique scatter matrices (parameters) Σ_x and Σ_y and positive constants τ_x^2 and τ_y^2 which satisfy

$$\mathbf{E}\left\{\mathbf{U}\left(\Sigma_x^{-1/2}(\mathbf{x}_1-\mathbf{x}_2)\right)\mathbf{U}\left(\Sigma_x^{-1/2}(\mathbf{x}_1-\mathbf{x}_3)\right)'\right\} = \tau_x^2\mathbf{I}_r$$

and

$$\mathbf{E}\left\{\mathbf{U}\left(\Sigma_y^{-1/2}(\mathbf{y}_1-\mathbf{y}_2)\right)\mathbf{U}\left(\Sigma_y^{-1/2}(\mathbf{y}_1-\mathbf{y}_3)\right)'\right\} = \tau_y^2\mathbf{I}_s,$$

respectively, and that $\hat{\Sigma}_x$ and $\hat{\Sigma}_y$ are root-n consistent estimates of Σ_x and Σ_y, again up to a constant. If we then choose

$$(\hat{\mathbf{R}}_{\mathbf{X}})_i = \mathbf{AVE}_j\left\{\mathbf{U}\left(\hat{\Sigma}_x^{-1/2}(\mathbf{x}_i-\mathbf{x}_j)\right)\right\} \text{ and } (\hat{\mathbf{R}}_{\mathbf{Y}})_i = \mathbf{AVE}_j\left\{\mathbf{U}\left(\hat{\Sigma}_y^{-1/2}(\mathbf{y}_i-\mathbf{y}_j)\right)\right\},$$

$i = 1,...,n$, one can show again as in Taskinen et al. (2005) that, under these mild assumptions and under the null hypothesis, $Q^2 \to_d \chi^2_{rs}$. (For our proposal above, one needs to show that the shape matrix estimate based on the spatial ranks is root-n consistent.)

Extension of Kendall's tau. As in the Spearman rho case, standardize the observation first ($\mathbf{x}_i \to \hat{\mathbf{x}}_i$ and $\mathbf{y}_i \to \hat{\mathbf{y}}_i$) so that

$$r\cdot \hat{\mathbf{R}}_{\mathbf{X}}'\hat{\mathbf{R}}_{\mathbf{X}} = |\hat{\mathbf{R}}_{\mathbf{X}}|^2\mathbf{I}_r \quad \text{and} \quad s\cdot \hat{\mathbf{R}}_{\mathbf{Y}}'\hat{\mathbf{R}}_{\mathbf{Y}} = |\hat{\mathbf{R}}_{\mathbf{Y}}|^2\mathbf{I}_s.$$

Then form a new data matrix of $n(n-1)$ differences

$$(\hat{\mathbf{X}},\hat{\mathbf{Y}})_d = (\hat{\mathbf{X}}_d,\hat{\mathbf{Y}}_d)$$

with rows consisting of all possible differences

$$((\hat{\mathbf{x}}_i-\hat{\mathbf{x}}_j)',(\hat{\mathbf{y}}_i-\hat{\mathbf{y}}_j)'), \quad i \neq j = 1,...,n.$$

Then transform to the spatial signs

$$(\hat{\mathbf{X}}, \hat{\mathbf{Y}})_s \to (\hat{\mathbf{U}}_{\mathbf{X}d}, \hat{\mathbf{U}}_{\mathbf{Y}d}).$$

The extension of Kendall's tau statistic

$$Q^2 = Q^2(\mathbf{X}, \mathbf{Y}) = nrs \frac{|\hat{\mathbf{U}}'_{\mathbf{X}d} \hat{\mathbf{U}}_{\mathbf{Y}d}|^2}{(n-1)^2 4 |\hat{\mathbf{R}}_{\mathbf{X}}|^2 |\hat{\mathbf{R}}_{\mathbf{Y}}|^2}$$

is then again asymptotically chi-square distributed with rs degrees of freedom. For details, see Taskinen et al. (2005).

10.4 Efficiency comparisons

To illustrate and compare the efficiencies of different test statistics for independence, we derive the limiting distributions of the test statistics under specific contiguous alternative sequences. Let $\mathbf{x}_i^*$ and $\mathbf{y}_i^*$ be independent with spherical marginal densities $\exp\{-\rho_x(|\mathbf{x}|)\}$ and $\exp\{-\rho_y(|\mathbf{y}|)\}$ and write, for some choices of $\mathbf{M}_1$ and $\mathbf{M}_2$,

$$\begin{pmatrix} \mathbf{x}_i \\ \mathbf{y}_i \end{pmatrix} = \begin{pmatrix} (1-\Delta)\mathbf{I}_r & \Delta \mathbf{M}_1 \\ \Delta \mathbf{M}_2 & (1-\Delta) I_s \end{pmatrix} \begin{pmatrix} \mathbf{x}_i^* \\ \mathbf{y}_i^* \end{pmatrix},$$

with $\Delta = \delta/\sqrt{n}$. Let f_Δ be the density of $(\mathbf{x}_i', \mathbf{y}_i')'$. Note that the joint distribution of $((\mathbf{x}_i^*)', (\mathbf{y}_i^*)')'$ is not spherically symmetric any more; the only exception is the multivariate normal case. The optimal likelihood ratio test statistic for testing H_0 against H_Δ is then

$$L = \sum_{i=1}^n \{\log f_\Delta(\mathbf{x}_i, \mathbf{y}_i) - \log f_0(\mathbf{x}_i, \mathbf{y}_i)\}.$$

We need the general assumption (which must be checked separately in each case) that, under the null hypothesis,

$$L = \frac{\delta}{\sqrt{n}} \sum_{i=1}^n \left[r - \psi_x(r_{xi}) r_{xi} + \psi_x(r_{xi}) r_{yi} \mathbf{u}'_{xi} \mathbf{M}_1 \mathbf{u}_{yi} \right.$$
$$\left. s - \psi_y(r_{yi}) r_{yi} + \psi_y(r_{yi}) r_{xi} \mathbf{u}'_{yi} \mathbf{M}_2 \mathbf{u}_{xi} \right] + o_P(1),$$

where $r_{xi} = |\mathbf{x}_i|$, $\mathbf{u}_{xi} = |\mathbf{x}_i|^{-1}\mathbf{x}_i$, and $\psi_x(r) = \rho_x'(r)$, and similarly for r_{yi}, $\mathbf{u}_{yi}$ and $\psi_y(r)$. Under this assumption the sequence of alternatives is contiguous to the null hypothesis.

Under the above sequence of alternatives we get the following limiting distributions.

Theorem 10.1. *Assume that* $max(p,q) > 1$. *The limiting distribution of the multivariate Blomqvist statistic under the sequence of alternatives is a noncentral chi-square distribution with rs degrees of freedom and noncentrality parameter*

$$\frac{\delta^2}{rs}|c_1\mathbf{M}_1 + c_2\mathbf{M}_2'|^2,$$

where

$$c_1 = (r-1)E(|\mathbf{y}_i^*|)E(|\mathbf{x}_i^*|^{-1}) \quad \text{and} \quad c_2 = (s-1)E(|\mathbf{x}_i^*|)E(|\mathbf{y}_i^*|^{-1}).$$

The limiting distribution of multivariate Spearman's rho and multivariate Kendall's tau under the sequence of alternatives above is a noncentral chi-square distribution with rs degrees of freedom and noncentrality parameter

$$\frac{\delta^2}{4\,rs\,\tau_x^2\tau_y^2}|d_1\mathbf{M}_1 + d_2\mathbf{M}_2'|^2,$$

where

$$d_1 = (r-1)E(|\mathbf{y}_i^* - \mathbf{y}_j^*|)E(|\mathbf{x}_i^* - \mathbf{x}_j^*|^{-1})$$

and

$$d_2 = (s-1)E(|\mathbf{x}_i^* - \mathbf{x}_j^*|)E(|\mathbf{y}_i^* - \mathbf{y}_j^*|^{-1}).$$

Finally, the limiting distribution of Wilk's test statistic $n \log W$ (and Pillai's trace statistic $Q^2 = nW^*$) under the sequence of alternatives is a noncentral chi-square distribution with rs degrees of freedom and noncentrality parameter

$$\delta^2\,|\mathbf{M}_1 + \mathbf{M}_2'|^2.$$

The above results are thus found in the case when the null marginal distributions are spherically symmetric. If the marginal distributions are elliptically symmetric, the efficiencies are of the same type $|h_1\mathbf{M}_1 + h_2\mathbf{M}_2'|^2$, where h_1 and h_2 depend on the marginal spherical distributions and the test used. If the marginal distributions are of the same dimension and the same type (which implies that $h_1 = h_2$) then the relative efficiencies do not depend on $\mathbf{M}_1$ and $\mathbf{M}_2$ at all. Note that, unfortunately, the tests are not unbiased (positive noncentrality parameter) for all alternative sequences.

As the limiting distributions are of the same type, χ^2_{rs}, the efficiency comparisons may be based on noncentrality parameters only. The efficiency comparisons are now made in the multivariate normal, t_5 and t_{10} distribution cases, and contaminated normal distribution cases. For simplicity, we assume that $\mathbf{M}_1 = \mathbf{M}_2'$. The resulting limiting efficiencies for selected dimensions are listed in Table 10.1. Note that because limiting multinormality of the regular covariance matrix holds if the fourth moments of the underlying distribution are finite, $n \log W$ has a limiting distribution only when $\nu \geq 3$. When the underlying distribution is multivariate normal ($\nu = \infty$), the Wilks test is naturally the best one, but Kendall's tau and Spearman's rho are very competitive with it. As the underlying population becomes heavy-tailed (ν gets smaller), then the rank-based tests are better than the Wilks test. The extension of the Blomqvist quadrant test is good for heavy-tailed distributions and for high dimensions. For more details, simulation studies and efficiencies under other distributions, see Taskinen et al. (2003, 2005).

Table 10.1 Asymptotical relative efficiencies of Kendall's tau and Spearman's rho (and the Blomqvist quadrant test in parentheses) as compared to the Wilks test at different r- and s-variate t distributions for selected $\nu = \nu_1 = \nu_2$

		r		
	s	2	5	10
	2	1.12	1.14	1.16
		(0.79)	(0.91)	(0.96)
$\nu = 5$	5		1.17	1.19
			(1.05)	(1.10)
	10			1.20
				(1.16)
	2	1.00	1.02	1.03
		(0.69)	(0.80)	(0.84)
$\nu = 10$	5		1.04	1.05
			(0.92)	(0.96)
	10			1.07
				(1.01)
	2	0.93	0.95	0.96
		(0.62)	(0.71)	(0.75)
$\nu = \infty$	5		0.96	0.97
			(0.82)	(0.86)
	10			0.98
				(0.91)

10.5 A real data example

The dataset considered in this example is a subset of hemodynamic data collected as a part of the LASERI study (English title: Cardiovascular risk in young Finns study) using whole-body impedance cardiography and plethysmographic blood pressure recordings from fingers. The measurements are on 223 healthy subjects between 26 and 42 years of age. The hemodynamic variables were recorded both in a supine position and during a passive head-up tilt on a motorized table. During that experiment the subject spent the first ten minutes in a supine position, then the motorized table was tilted to a head-up position (60 degrees) for five minutes, and for the last five minutes the table was again returned to the supine position. We are interested in the differences between some recorded values before and after the tilt, denoted by HRT1T4, COT1T4, and SVRIT1T4. It is of interest whether these variables and the weight, height, and hip measurements are independent. The observed values can be seen in Figure 10.1.

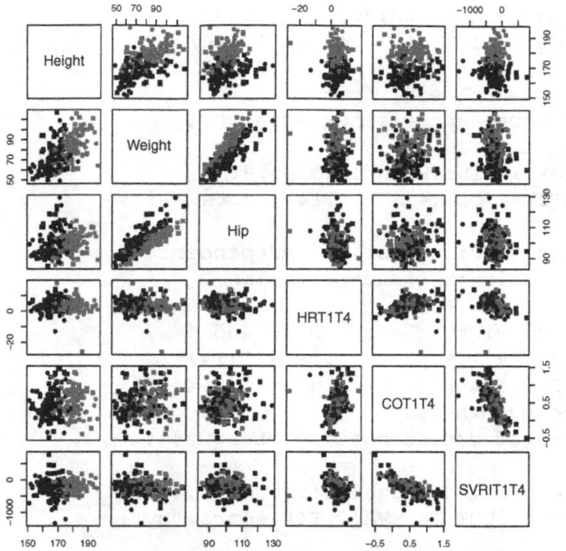

Fig. 10.1 A scatterplot matrix for LASERI data.

The figure is obtained using

```
> data(LASERI)
> attach(LASERI)
> pairs( cbind(Height,Weight,Hip,HRT1T4,COT1T4,SVRIT1T4),
  col=as.numeric(Sex), pch=15:16)
```

For the two 3-variate observations we first used the tests based on the identity score (Pillai's trace) and the tests based on standardized spatial signs of the observation vectors (an extension of the Blomqvist test) and those of the differences of the observation vectors (an extension of Kendall's tau). All the p-values are small; see the results below.

```
> mv.ind.test(cbind(Height,Waist,Hip),
  cbind(HRT1T4,COT1T4,SVRIT1T4))

        Test of independence using Pillai's trace
```

```
data:   cbind(Height, Waist, Hip) and
        cbind(HRT1T4, COT1T4, SVRIT1T4)
Q.2 = 20.371, df = 9, p-value = 0.01576
alternative hypothesis: true
measure of dependence is not equal to 0

> mv.ind.test(cbind(Height,Waist,Hip),
  cbind(HRT1T4,COT1T4,SVRIT1T4),score="si")

        Spatial sign test of independence
        using inner standardization

data:   cbind(Height, Waist, Hip) and
        cbind(HRT1T4, COT1T4, SVRIT1T4)
Q.2 = 24.1245, df = 9, p-value = 0.004109
alternative hypothesis:
true measure of dependence is not equal to 0

> mv.ind.test(cbind(Height,Waist,Hip),
  cbind(HRT1T4,COT1T4,SVRIT1T4),score="sy")

        Symmetrized spatial sign test
        of independence using inner
        standardization

data:   cbind(Height, Waist, Hip) and
        cbind(HRT1T4, COT1T4, SVRIT1T4)
Q.2 = 25.5761, df = 9, p-value = 0.002396
alternative hypothesis:
true measure of dependence is not equal to 0
```

We next dropped two variables Hip and COT1T4. For the remaining two bivariate observations we cannot reject the null hypothesis of independence, and one gets the following results (Pillai's trace and an extension of Spearman's rho; approximate p-values coming from a permutation tests are also given).

```
> mv.ind.test(cbind(Height,Waist),
  cbind(HRT1T4,SVRIT1T4))

        Test of independence using
        Pillai's trace

data:   cbind(Height, Waist) and
        cbind(HRT1T4, SVRIT1T4)
Q.2 = 3.9562, df = 4, p-value = 0.412
alternative hypothesis: true measure
of dependence is not equal to 0
```

```
> mv.ind.test(cbind(Height,Waist),
  cbind(HRT1T4,SVRIT1T4), method="p")

        Test of independence using
        Pillai's trace

data:  cbind(Height, Waist) and
       cbind(HRT1T4, SVRIT1T4)
Q.2 = 3.9562, replications = 1000, p-value = 0.422
alternative hypothesis:
true measure of dependence is not equal to 0

> mv.ind.test(cbind(Height,Waist),
  cbind(HRT1T4,SVRIT1T4), score="r", method="p")

        Spatial rank test of independence
        using inner standardization

data:  cbind(Height, Waist) and
       cbind(HRT1T4, SVRIT1T4)
Q.2 = 4.0556, replications = 1000, p-value = 0.408
alternative hypothesis:
true measure of dependence is not equal to 0

> mv.ind.test(cbind(Height,Waist),
  cbind(HRT1T4,SVRIT1T4), score="r", method="p")

        Spatial rank test of independence
        using inner standardization

data:  cbind(Height, Waist) and
       cbind(HRT1T4, SVRIT1T4)
Q.2 = 4.0556, replications = 1000, p-value = 0.408
alternative hypothesis:
true measure of dependence is not equal to 0
```

10.6 Canonical correlation analysis

Assume that $r \le s$. The classical canonical correlation analysis based on the covariance matrix finds the transformation matrices $\mathbf{H}_x$ and $\mathbf{H}_y$ such that

$$\mathbf{COV}(\mathbf{X}\mathbf{H}_x', \mathbf{Y}\mathbf{H}_y') = \begin{pmatrix} \mathbf{I}_r & \mathbf{L} \\ \mathbf{L}' & \mathbf{I}_s \end{pmatrix},$$

where $\mathbf{L} = (\mathbf{L}_0, \mathbf{O})$ with an $r \times r$ diagonal matrix $\mathbf{L}_0$. Note that Pillai's trace statistic is the sum of squared canonical correlations

$$W^* = tr(\mathbf{LL}') = tr(\mathbf{L}'\mathbf{L}).$$

Write again

$$\mathbf{COV}(\mathbf{X},\mathbf{Y}) = \mathbf{COV} = \begin{pmatrix} \mathbf{COV}_{11} & \mathbf{COV}_{12} \\ \mathbf{COV}_{21} & \mathbf{COV}_{22} \end{pmatrix}.$$

Then

$$\begin{aligned} \mathbf{H}_x\, \mathbf{COV}_{11}\, \mathbf{H}_x' &= \mathbf{I}_r, \\ \mathbf{H}_y\, \mathbf{COV}_{22}\, \mathbf{H}_y' &= \mathbf{I}_s, \text{ and} \\ \mathbf{H}_x\, \mathbf{COV}_{12}\, \mathbf{H}_y' &= \mathbf{L}. \end{aligned}$$

To simplify the notations, assume that $r = s$ and that the limiting distribution of (vectorized)

$$\sqrt{n}\left(\begin{pmatrix} \mathbf{COV}_{11} & \mathbf{COV}_{12} \\ \mathbf{COV}_{21} & \mathbf{COV}_{22} \end{pmatrix} - \begin{pmatrix} \mathbf{I}_r & \Lambda \\ \Lambda & \mathbf{I}_r \end{pmatrix}\right)$$

is a multivariate normal distribution with zero mean vector and the diagonal matrix Λ has distinct diagonal elements. Then, using the three equations above and Slutsky's theorem, also $\sqrt{n}(\mathbf{H}_x - \mathbf{I}_r)$, $\sqrt{n}(\mathbf{H}_y - \mathbf{I}_r)$, and $\sqrt{n}(\mathbf{L} - \Lambda)$ have multivariate normal limiting distributions that can be solved from

$$\begin{aligned} \sqrt{n}(\mathbf{H}_x - \mathbf{I}_r) + \sqrt{n}(\mathbf{H}_x' - \mathbf{I}_r) + \sqrt{n}(\mathbf{COV}_{11} - \mathbf{I}_r) &= o_P(1), \\ \sqrt{n}(\mathbf{H}_y - \mathbf{I}_r) + \sqrt{n}(\mathbf{H}_y' - \mathbf{I}_r) + \sqrt{n}(\mathbf{COV}_{22} - \mathbf{I}_r) &= o_P(1) \text{ and} \\ \sqrt{n}(\mathbf{H}_x - \mathbf{I}_r)\Lambda + \sqrt{n}\Lambda(\mathbf{H}_y' - \mathbf{I}_r) + \sqrt{n}(\mathbf{COV}_{12} - \Lambda) &= \sqrt{n}(\mathbf{L} - \Lambda) + o_P(1). \end{aligned}$$

In the multivariate normal case, the limiting distribution of standardized canonical correlations $\sqrt{n}(l_{ii} - \Lambda_{ii})$, $i = 1, \ldots, r$, for example, is $N(0, (1 - \Lambda_{ii}^2)^2)$, and the canonical correlations are asymptotically independent. See Anderson (1999). Naturally, in the general elliptic case, any root-n consistent and asymptotically normal scatter matrix $\mathbf{S}$ can be used for the canonical analysis in a similar way. Also in that case the limiting distribution of standardized canonical correlations $\sqrt{n}(l_{ii} - \Lambda_{ii})$, $i = 1, \ldots, r$, is a multivariate normal distribution $N(0, (1 - \Lambda_{ii}^2)^2 \mathrm{ASV}(S_{12}))$, where $\mathrm{ASV}(S_{12})$ is the limiting variance of the off-diagonal element of $\mathbf{S}(\mathbf{X})$ when $\mathbf{X}$ is a random sample from a corresponding $(r+s)$-variate spherical distribution F standardized so that $\mathbf{S}(F) = \mathbf{I}_{r+s}$. Also, the canonical correlations are asymptotically independent. See Taskinen et al. (2006) for details.

10.7 Other approaches

A nonparametric analogue to Wilks' test was given by Puri and Sen (1971). They developed a class of tests based on componentwise ranking which uses a test statistic of the form

$$S^J = \frac{|T|}{|T_{11}||T_{22}|}.$$

Here the elements of $(r+s) \times (r+s)$ matrix T are

$$T_{kl} = \frac{1}{n} \sum_{i=1}^{n} J\left(\frac{R_{ki}}{n+1}\right) J\left(\frac{R_{li}}{n+1}\right),$$

where R_{ki} denotes the rank of the kth component of $(\mathbf{x}_i', \mathbf{y}_i')'$ among the kth components of all n vectors and $J(\cdot)$ is an arbitrary (standardized) score function. Under H_0, $-n \log S^J \to_d \chi^2_{rs}$.

Gieser and Randles (1997) proposed a nonparametric test based on interdirection counts that generalized the univariate $(p = q = 1)$ quadrant test of Blomqvist (1950) and which is invariant under the transformation group. Taskinen et al. (2003) gave a more practical invariant extension of the quadrant test based on spatial signs. It is easy to compute for data in common dimensions. These two tests are asymptotically equivalent if the marginal vectors are elliptically distributed. Later Taskinen et al. (2005) introduced multivariate extensions of Kendall's tau and Spearman's rho that are based on interdirections. Sukhanova et al. (2009) gave extensions based on the Oja signs and ranks.

Oja et al. (2009) found optimal nonparametric tests of independence in the symmetric independent component model. These tests were based on marginal signed-ranks applied to the (estimated) independent components.

Chapter 11
Several-sample location problem

Abstract In this chapter we consider tests and estimates based on identity, spatial sign, and spatial rank scores in the several independent samples setting. We get multivariate extensions of the Moods test, Wilcoxon-Mann-Whitney test, Kruskal-Wallis test and the two samples Hodges-Lehmann estimator. Equivariant/invariant versions are found using inner centering and standardization.

11.1 A general strategy for testing and estimation

Let now data matrix

$$\mathbf{Y} = (\mathbf{Y}_1', ..., \mathbf{Y}_c')'$$

consist of c independent random samples

$$\mathbf{Y}_i = (\mathbf{y}_{i1}, ..., \mathbf{y}_{in_i})', \quad i = 1, ..., c,$$

from p-variate distributions with cumulative distribution functions $F_1, ..., F_c$. We write $n = n_1 + \cdots + n_c$ for the total sample size. We assume that the independent p-variate observation vectors $\mathbf{y}_{ij}$ with cumulative distribution functions F_i are generated by

$$\mathbf{y}_{ij} = \mu_i + \Omega\, \varepsilon_{ij}, \quad i = 1, ..., c; \; j = 1, ..., n_i,$$

where the ε_{ij} are independent and standardized vectors all having the same unknown distribution. Then $\mu_1, ..., \mu_c$ are unknown location centers and the matrix $\Sigma = \Omega\Omega' > 0$ is a joint unknown scatter matrix specified later. At first, we wish to test the null hypothesis

$$H_0: \; \mu_1 = \cdots = \mu_c.$$

Under our assumptions, the null hypothesis can also be written as $H_0: \; F_1 = \cdots = F_c$ saying that all observations come from the same population (back in the one-sample case). Later in this chapter, we also wish to estimate unknown centers $\mu_1, ..., \mu_c$ or differences $\Delta_{ij} = \mu_j - \mu_i$, $i, j = 1, ..., c$.

H. Oja, *Multivariate Nonparametric Methods with R: An Approach Based on Spatial Signs and Ranks*, Lecture Notes in Statistics 199, DOI 10.1007/978-1-4419-0468-3_11,

Again, as in the one-sample case, we wish to use a general location score function $\mathbf{T}(\mathbf{y})$ in testing and estimation. Using inner centering and outer standardization, the procedure for testing proceeds as follows.

Test statistic using inner centering and outer standardization.

1. Find a shift vector $\hat{\mu}$ such that

$$\mathbf{AVE}\{\mathbf{T}(\mathbf{y}_{ij} - \hat{\mu})\} = \mathbf{0}.$$

For the inner centered scores, write

$$\hat{\mathbf{T}}_{ij} = \mathbf{T}(\mathbf{x}_{ij} - \hat{\mu}), \quad i = 1,...,c; \; j = 1,...,n_i.$$

2. The test statistic for testing whether the ith sample differs from the others is then based on

$$\hat{\mathbf{T}}_i = \mathbf{AVE}_j\left\{\hat{\mathbf{T}}_{ij}\right\}, \quad i = 1,...,c.$$

3. The test statistic for $H_0 : F_1 = \cdots = F_c$ is

$$Q^2 = Q^2(\mathbf{Y}) = \sum_{i=1}^{c} \left\{n_i \hat{\mathbf{T}}_i' \hat{\mathbf{B}}^{-1} \hat{\mathbf{T}}_i\right\},$$

where

$$\hat{\mathbf{B}} = \mathbf{AVE}\left\{\hat{\mathbf{T}}_{ij}\hat{\mathbf{T}}_{ij}'\right\}.$$

4. Under the null hypothesis $H_0 : \mu_1 = \cdots = \mu_c$ and under assumptions specified later, the limiting distribution of $Q^2(\mathbf{Y})$ is $\chi^2_{(c-1)p}$.

Inner centering makes the test statistic location invariant; that is,

$$Q^2(\mathbf{Y} + \mathbf{1}_n \mathbf{b}') = Q^2(\mathbf{Y})$$

for all p-vectors $\mathbf{b}$. Note that the test statistic can be written as

$$Q^2 = tr\left(\sum_{i=1}^{c} \{n_i \hat{\mathbf{T}}_i \hat{\mathbf{T}}_i'\} \, \mathbf{AVE}\left\{\hat{\mathbf{T}}_{ij}\hat{\mathbf{T}}_{ij}'\right\}^{-1}\right)$$

which compares two scatter matrices (for "between" and "total" variation). In fact Q^2 is the classical Pillai trace statistic for MANOVA, now based on centered score values $\hat{\mathbf{T}}_{ij}$ instead of original centered values $\mathbf{y}_{ij} - \bar{\mathbf{y}}$.

The approach based on inner centering and inner standardization is given next.

Test statistic using inner centering and inner standardization.

1. Find a shift vector $\hat{\mu}$ and full rank transformation matrix $\mathbf{S}^{-1/2}$ such that the scores

$$\hat{\mathbf{T}}_{ij} = \mathbf{T}(\mathbf{S}^{-1/2}(\mathbf{y}_{ij} - \hat{\mu})), \quad i = 1,...,c; \ j = 1,...,n_i,$$

are standardized and centered in the sense that

$$\mathbf{AVE}\left\{\hat{\mathbf{T}}_{ij}\right\} = \mathbf{0} \quad \text{and} \quad p \cdot \mathbf{AVE}\left\{\hat{\mathbf{T}}_{ij}\hat{\mathbf{T}}_{ij}'\right\} = \mathbf{AVE}\left\{|\hat{\mathbf{T}}_{ij}|^2\right\}\mathbf{I}_p.$$

This is the inner centering and inner standardization.

2. The test statistic for testing whether the ith sample differs from the others is then based on

$$\hat{\mathbf{T}}_i = \mathbf{AVE}_j\left\{\hat{\mathbf{T}}_{ij}\right\}, \quad i = 1,...,n.$$

3. The several-sample location test statistic is then (under some assumptions)

$$Q^2 = np\frac{\sum_i n_i |\hat{\mathbf{T}}_i|^2}{\sum_i \sum_j |\hat{\mathbf{T}}_{ij}|^2}.$$

4. Under the null hypothesis $H_0: \mu_1 = \cdots = \mu_c$ and under assumptions specified later, the limiting distribution of $Q^2(\mathbf{Y})$ is $\chi^2_{(c-1)p}$.

The test using both inner centering and inner standardization is affine invariant; that is,

$$Q^2(\mathbf{Y}\mathbf{H}' + \mathbf{1}_n\mathbf{b}') = Q^2(\mathbf{Y})$$

for all full-rank $p \times p$ matrices $\mathbf{H}$ and for all p-vectors $\mathbf{b}$. For both test versions, under general assumptions stated later, the limiting distribution of the test statistic Q^2 is a chi-square distribution with $(c-1)p$ degrees of freedom that can be used to calculate approximate p-values. The p-value can also be calculated for the conditionally distribution-free, exact *permutation test* version. Let $\mathbf{P}$ be an $n \times n$ permutation matrix (obtained from an identity matrix by permuting rows or columns). The p-value of the permutation test statistic is then

$$\mathbf{E}_\mathbf{P}\left[I\left(Q^2(\mathbf{P}\mathbf{Y}) \geq Q^2(\mathbf{Y})\right)\right],$$

where $\mathbf{P}$ has a uniform distribution over all possible $n!$ permutations.

11.2 Hotelling's T^2 and MANOVA

We start with the classical multivariate analysis of variance (MANOVA) test statistics which are given by the choice $\mathbf{T}(\mathbf{y}) = \mathbf{y}$, identity score function. Again, write

$$\mathbf{Y} = (\mathbf{Y}_1', ..., \mathbf{Y}_c')' \quad \text{and} \quad \mathbf{Y}_i = (\mathbf{y}_{i1}, ..., \mathbf{y}_{in_i})', \quad i = 1,...,c,$$

and let the observations $\mathbf{y}_{ij}$ be generated by

$$\mathbf{y}_{ij} = \mu_i + \Omega\varepsilon_{ij}, \quad i = 1,...,c; \ j = 1,...,n_i,$$

where ε_{ij} are independent, centered, and standardized observations with cumulative distribution function F. In this first approach we assume that the second moments of ε_{ij} exist and

$$\mathbf{E}(\varepsilon_{ij}) = \mathbf{0} \quad \text{and} \quad \mathbf{COV}(\varepsilon_{ij}) = \mathbf{I}_p.$$

Then $\mathbf{E}(\mathbf{y}_{ij}) = \mu_i$ and $\mathbf{COV}(\mathbf{y}_{ij}) = \Sigma = \Omega\Omega'$. First, we wish to test the null hypothesis

$$H_0: \ \mu_1 = \cdots = \mu_c.$$

Our general testing strategy with the identity score function then uses sample mean vectors

$$\bar{\mathbf{y}}_i = \mathbf{AVE}_j\{\mathbf{y}_{ij}\}, \quad i = 1, ..., c,$$

"grand mean vector"

$$\bar{\mathbf{y}} = \mathbf{AVE}\{\mathbf{y}_{ij}\},$$

and sample covariance matrix

$$\hat{\mathbf{B}} = \mathbf{S} = \mathbf{AVE}\{(\mathbf{y}_{ij} - \bar{\mathbf{y}})(\mathbf{y}_{ij} - \bar{\mathbf{y}})'\}.$$

Then we get the following.

Test statistic using inner centering and outer standardization.

1. Inner (and outer) centered scores are
$$\hat{\mathbf{T}}_{ij} = \mathbf{y}_{ij} - \bar{\mathbf{y}}, \quad i = 1, ..., c; \ j = 1, ..., n_i.$$
2. The test statistic for testing whether the ith sample differs from the others is then based on
$$\hat{\mathbf{T}}_i = \bar{\mathbf{y}}_i - \bar{\mathbf{y}}, \quad i = 1, ..., c.$$
3. The test statistic is
$$Q^2 = Q^2(\mathbf{Y}) = \sum_{i=1}^{c} \left\{ n_i (\bar{\mathbf{y}}_i - \bar{\mathbf{y}})' \mathbf{S}^{-1} (\bar{\mathbf{y}}_i - \bar{\mathbf{y}}) \right\}.$$
4. Under the null hypothesis $H_0: \ \mu_1 = \cdots = \mu_c$, the limiting distribution of $Q^2(\mathbf{Y})$ is $\chi^2_{(c-1)p}$.

It is straightforward to see that the inner centering and inner standardization yield exactly the same test statistic. The classical MANOVA procedure usually starts with a decomposition

$$SS_T = SS_B + SS_W$$

corresponding to

$$\sum_i \sum_j (\mathbf{y}_{ij} - \bar{\mathbf{y}})(\mathbf{y}_{ij} - \bar{\mathbf{y}})' = \sum_i \sum_j (\bar{\mathbf{y}}_i - \bar{\mathbf{y}})(\bar{\mathbf{y}}_i - \bar{\mathbf{y}})' + \sum_i \sum_j (\mathbf{y}_{ij} - \bar{\mathbf{y}}_i)(\mathbf{y}_{ij} - \bar{\mathbf{y}}_i)'.$$

Thus the "total" variation SS_T is decomposed into a sum of "between" and "within" variations, SS_B and SS_W. Our test statistic

$$Q^2 = n \cdot tr\left(SS_B SS_T^{-1}\right)$$

compares the "between" and "total" matrices and it is known in the literature with the name *Pillai's trace* statistic. Another possibility is to base the test on the *Lawley-Hotelling's trace* statistic, $Q_{LH}^2 = n \cdot tr(SS_B SS_W^{-1})$. The test statistics Q^2 (correspondingly Q_{LH}^2) is simply n times the sum of eigenvalues of $SS_B SS_T^{-1}$ (correspondingly $SS_B SS_W^{-1}$). Instead of considering the sum (or the arithmetic mean) of the eigenvalues of $SS_B SS_T^{-1}$ or $SS_B SS_W^{-1}$, one could base the test on the product (or the geometrical mean) of the eigenvalues. The so-called *Wilks' lambda*, for example, is $\Lambda = \det(S_W SS_T^{-1})$. In fact, Wilks' lambda is the likelihood ratio test statistic in the multivariate normal case.

Another formulation of the problem and test statistics. In the practical analysis of data, the data are usually given in the form

$$(\mathbf{X}, \mathbf{Y}),$$

where $\mathbf{X}$ is an $n \times c$ matrix indicating the group (sample) membership and the $n \times p$ matrix $\mathbf{Y}$ gives the values of the p-variate response vector. Then

$$\mathbf{X}_{ij} = \begin{cases} 1, & \text{if the } i\text{th observation comes from group (sample) } j \\ 0, & \text{otherwise.} \end{cases}$$

Matrices $\mathbf{XX}'$ and $\mathbf{X}'\mathbf{X}$ have nice interpretations, namely

$$(\mathbf{XX}')_{ij} = \begin{cases} 1, & \text{if } i\text{th and } j\text{th observations come from the same treatment group} \\ 0, & \text{otherwise.} \end{cases}$$

and $\mathbf{X}'\mathbf{X}$ is a $c \times c$ diagonal matrix whose diagonal elements are the c group sizes, here denoted by $n_1, \ldots, n_c$.

For the regular outer centering of the observation vectors, one can use the projection matrix

$$\mathbf{P}_{\mathbf{1}_n} = \mathbf{1}_n(\mathbf{1}_n'\mathbf{1}_n)^{-1}\mathbf{1}_n' = \frac{1}{n}\mathbf{1}_n\mathbf{1}_n'.$$

The projection $\mathbf{Y} \rightarrow \mathbf{P}_{\mathbf{1}_n}\mathbf{Y}$ then replaces the observations by the "grand" mean vector, and the outer (and inner) centered observations are obtained as residuals,

$$\mathbf{Y} \rightarrow \hat{\mathbf{Y}} = (\mathbf{I}_n - \mathbf{P}_{\mathbf{1}_n})\mathbf{Y} = \left(\mathbf{I}_n - \frac{1}{n}\mathbf{1}_n\mathbf{1}_n'\right)\mathbf{Y}.$$

Next recall that

$$\mathbf{P_X} = \mathbf{X}(\mathbf{X}'\mathbf{X})^{-1}\mathbf{X}'$$

is the projection matrix that projects the data points to the subspace spanned by the columns of $\mathbf{X}$. This means that, in our case, transformation $\mathbf{Y} \to \mathbf{P_X}\mathbf{Y}$ replaces the observation vectors by their group mean vectors. Matrix $\mathbf{I}_n - \mathbf{P_X}$ is the projection matrix to the corresponding residual space; that is, matrix $(\mathbf{I}_n - \mathbf{P_X})\mathbf{Y}$ yields the differences between observations and the corresponding group (sample) means.

Using these notations,

$$\hat{\mathbf{Y}}'\hat{\mathbf{Y}} = \hat{\mathbf{Y}}'\mathbf{P_X}\hat{\mathbf{Y}} + \hat{\mathbf{Y}}'(\mathbf{I}_n - \mathbf{P_X})\hat{\mathbf{Y}},$$

and therefore

$$Q^2 = n \cdot tr\left(\hat{\mathbf{Y}}'\mathbf{P_X}\hat{\mathbf{Y}}(\hat{\mathbf{Y}}'\hat{\mathbf{Y}})^{-1}\right)$$

and

$$Q^2_{LH} = n \cdot tr\left(\hat{\mathbf{Y}}'\mathbf{P_X}\hat{\mathbf{Y}}(\hat{\mathbf{Y}}'(\mathbf{I}_n - \mathbf{P_X})\hat{\mathbf{Y}})^{-1}\right).$$

Recall that in the one-sample case, for testing $H_0: \mu = \mathbf{0}$, one uses the decomposition

$$\mathbf{Y}'\mathbf{Y} = \mathbf{Y}'\mathbf{P}_{\mathbf{1}_n}\mathbf{Y} + \mathbf{Y}'(\mathbf{I}_n - \mathbf{P}_{\mathbf{1}_n})\mathbf{Y} = \mathbf{Y}'\mathbf{P}_{\mathbf{1}_n}\mathbf{Y} + \hat{\mathbf{Y}}'\hat{\mathbf{Y}}$$

and there we obtained

$$Q^2 = n \cdot tr(\mathbf{Y}'\mathbf{P}_{\mathbf{1}_n}\mathbf{Y}(\mathbf{Y}'\mathbf{Y})^{-1}) \quad \text{and} \quad T^2 = n \cdot tr(\mathbf{Y}'\mathbf{P}_{\mathbf{1}_n}\mathbf{Y}(\hat{\mathbf{Y}}'\hat{\mathbf{Y}})^{-1}).$$

Model of multivariate normality. We assume first that $\mathbf{Y}_i$ is a random sample from a multivariate normal distribution $N_p(\mu_i, \Sigma)$, $i = 1, \ldots, c$. If $c = 2$,

$$Q^2 = \frac{n_1 n_2}{n}(\bar{\mathbf{y}}_1 - \bar{\mathbf{y}}_2)'\mathbf{S}^{-1}(\bar{\mathbf{y}}_1 - \bar{\mathbf{y}}_2),$$

similarly Q^2_{LH}, are different versions of *Hotelling's two-sample T^2-test.* In the general c-sample case, Q^2_{LH} is the classical MANOVA test statistic, namely the Lawley-Hotelling trace statistic. The test statistic Q^2 is known as Pillai's trace statistic. The exact distributions of the test statistics are known but quite complicated; see, for example, DasGupta (1999a,b) for a discussion of that. In practice, one can often use the approximate distribution which is a chi-square distribution with $(c-1)p$ degrees of freedom. This is discussed next.

Nonparametric model with finite second moments. The test statistics Q^2 and Q^2_{LH} are also asymptotically valid in a wider nonparametric model where we only assume that the second moments exist. In this model, we then assume that the observations are generated according to

$$\mathbf{y}_{ij} = \mu_i + \Omega\,\varepsilon_{ij}, \quad i = 1, \ldots, c; \; j = 1, \ldots, n_i,$$

where ε_{ij} are independent vectors all having the same unknown distribution with $\mathbf{E}(\varepsilon_{ij}) = \mathbf{0}$ and $\mathbf{COV}(\varepsilon_{ij}) = \mathbf{I}_p$. Recall that the vectors $\mu_1, ..., \mu_c$ are unknown population mean vectors, and $\Sigma = \Omega\Omega'$ is the covariance matrix of $\mathbf{y}_{ij}$.

Next we consider the limiting null distribution of the test statistic. For that we need the assumption that

$$\frac{n_i}{n} \to \lambda_i, \quad i = 1, ..., c, \quad \text{as } n \to \infty,$$

where $0 < \lambda_i < 1$, $i = 1, ..., c$. First, under the null hypothesis,

$$\frac{1}{n}\hat{\mathbf{Y}}'\hat{\mathbf{Y}} \to_P \Sigma \quad \text{and} \quad \frac{1}{n}\hat{\mathbf{Y}}'(\mathbf{I}_n - \mathbf{P_X})\hat{\mathbf{Y}} \to_P \Sigma$$

and therefore Q^2_{LH} and Q^2 are asymptotically equivalent; that is, $Q^2_{LH} - Q^2 \to_P 0$. Then note, using the multivariate central limit theorem (CLT), that the limiting distribution of

$$(\sqrt{n_1}\bar{\mathbf{y}}_1', ..., \sqrt{n_{c-1}}\bar{\mathbf{y}}_{c-1}')'$$

is a multivariate normal distribution with mean vector zero and covariance matrix

$$\left(\mathbf{I}_{c-1} - \mathbf{dd}'\right) \otimes \Sigma,$$

where $\mathbf{d} = (\sqrt{\lambda_1}, ..., \sqrt{\lambda_{c-1}})'$. We use the first $c-1$ sample mean vectors only as the covariance matrix of $(\sqrt{n_1}\bar{\mathbf{y}}_1', ..., \sqrt{n_c}\bar{\mathbf{y}}_c')'$ is singular. Then we find that

$$\begin{aligned} Q^2 &= \mathbf{AVE}_i \left\{ n_i \bar{\mathbf{y}}_i' \mathbf{S}^{-1} \bar{\mathbf{y}}_i \right\} \\ &= (\sqrt{n_1}\bar{\mathbf{y}}_1', ..., \sqrt{n_{c-1}}\bar{\mathbf{y}}_{c-1}') \left(\left(\mathbf{I}_{c-1} - \hat{\mathbf{d}}\hat{\mathbf{d}}'\right)^{-1} \otimes \mathbf{S}^{-1} \right) \begin{pmatrix} \sqrt{n_1}\bar{\mathbf{y}}_1 \\ \cdots \\ \sqrt{n_{c-1}}\bar{\mathbf{y}}_{c-1} \end{pmatrix} \end{aligned}$$

with $\hat{d}_i = \sqrt{n_i/n}$ and with

$$\left(\mathbf{I}_{c-1} - \hat{\mathbf{d}}\hat{\mathbf{d}}'\right)^{-1} = \left(\mathbf{I}_{c-1} + \frac{\hat{\mathbf{d}}\hat{\mathbf{d}}'}{1 - \hat{\mathbf{d}}'\hat{\mathbf{d}}}\right).$$

This shows that the limiting distribution of Q^2 is a chi-square distribution with $(c-1)p$ degrees of freedom. Approximate power is obtained if one considers a sequence of alternatives $H_n: \ \mu_i = \mu + n^{-1/2}\delta_i$, $i = 1, ..., c$, where $\sum_{i=1}^c \lambda_i \delta_i = 0$. It is straightforward to see that, under this sequence of alternatives, the limiting distribution of Q^2 is a noncentral chi square distribution with $(c-1)p$ degrees of freedom and the noncentrality parameter

$$\sum_{i=1}^{c} \lambda_i \delta_i' \Sigma^{-1} \delta_i.$$

We have thus proved the following.

Theorem 11.1.

- *Under assumptions stated above and under the null hypothesis $H_0: \mu_1 = \cdots = \mu_c$, the limiting distribution of Q^2 is a chi-square distribution with $(c-1)p$ degrees of freedom.*
- *Under assumptions stated above and under the sequence of alternatives H_n : $\mu_i = \mu + n^{-1/2}\delta_i$, $i = 1,...,c$, where $\sum_{i=1}^{c} \lambda_i \delta_i = 0$, the limiting distribution of the test statistic Q^2 is a chi-square distribution with $(c-1)p$ degrees of freedom and a noncentrality parameter $\sum_{i=1}^{c} \lambda_i \delta_i' \Sigma^{-1} \delta_i$.*

Assume that $\mathbf{E}(\mathbf{Y}) = \Delta$ and center the mean matrix as $\hat{\Delta} = (\mathbf{I}_n - \mathbf{P}_{\mathbf{1}_n})\Delta$. According to Theorem 11.1, the distribution of Q^2 can be approximated by a noncentral chi-square distribution with $(c-1)p$ degrees of freedom and noncentrality parameter

$$tr(\hat{\Delta}' \mathbf{P}_\mathbf{X} \hat{\Delta} \Sigma^{-1}).$$

This result can be used in the power and sample size calculations.

Totally nonparametric model. The permutation version of the MANOVA test is based on the fact that under the null distribution

$$\mathbf{PY} \sim \mathbf{Y}$$

for all $n \times n$ permutation matrices $\mathbf{P}$. In this approach we use the second formulation of the model. Now the value of the test statistic for permuted data set $\mathbf{PY}$ is

$$Q^2(\mathbf{PY}) = n \cdot tr\left(\hat{\mathbf{Y}}'\mathbf{P}\mathbf{P}_\mathbf{X}\mathbf{P}\hat{\mathbf{Y}}(\hat{\mathbf{Y}}'\hat{\mathbf{Y}})^{-1}\right).$$

To calculate a new value of the test statistic after the permutation, we only need to transform the projection matrix $\mathbf{P}_\mathbf{X} \rightarrow \mathbf{P}\mathbf{P}_\mathbf{X}\mathbf{P}$. The exact p value is again obtained as

$$E_\mathbf{P}\left\{Q^2(\mathbf{PY}) \geq Q^2(\mathbf{Y})\right\},$$

where $\mathbf{P}$ is uniformly distributed in the set of all different $n!$ permutation matrices.

Some final comments. It is easy to check that the test statistic Q^2 is *affine invariant* in the sense that

$$Q^2(\mathbf{YH}' + \mathbf{1}_n\mathbf{b}') = Q^2(\mathbf{Y})$$

for any full-rank $p \times p$ transformation matrix $\mathbf{H}$ and for any p-vector $\mathbf{b}$. This is naturally true for Q^2_{LH} as well. This means that the exact null distribution then does not depend on μ or Ω at all. If we write

$$\hat{\varepsilon} = \hat{\mathbf{Y}}\mathbf{S}^{-1/2} = (\mathbf{I}_n - \mathbf{P}_{\mathbf{1}_n})\mathbf{Y}\mathbf{S}^{-1/2}$$

then the test statistic simplifies to

$$Q^2(\mathbf{Y}) = Q^2(\hat{\varepsilon}) = tr(\hat{\varepsilon}'\mathbf{P_X}\hat{\varepsilon}) = |\mathbf{P}_X\hat{\varepsilon}|^2,$$

which is just the squared norm of a projection of the centered and standardized data matrix.

11.3 The test based on spatial signs

In this section we consider the test statistic that uses the spatial sign score function $\mathbf{T}(\mathbf{y}) = \mathbf{U}(\mathbf{y})$. The tests are then extensions of univariate two- and several-sample *Mood's test* or *median test*; see Mood (1954). Again, we assume that the observations $\mathbf{y}_{ij}$ are generated by

$$\mathbf{y}_{ij} = \mu_i + \Omega\varepsilon_{ij}, \quad i = 1,...,c; \; j = 1,...,n_i,$$

where ε_{ij} are independent, centered, and standardized observations (in the sense described later) with a joint cumulative distribution function F. Again, we wish to test the null hypothesis

$$H_0: \; \mu_1 = \cdots = \mu_c.$$

Then the first testing procedure is as follows.

Test statistic using inner centering and outer standardization.

1. Inner centered scores are

$$\hat{\mathbf{U}}_{ij} = \mathbf{U}(\mathbf{y}_{ij} - \hat{\mu}), \quad i = 1,...,c; \; j = 1,...,n_i,$$

   where $\hat{\mu} = \hat{\mu}(\mathbf{Y})$ is the spatial median so that $\mathbf{AVE}\{\hat{\mathbf{U}}_{ij}\} = \mathbf{0}$.
2. The test statistic for testing whether the ith sample differs from the others is based on

$$\hat{\mathbf{U}}_i = \mathbf{AVE}_j\{\hat{\mathbf{U}}_{ij}\}, \quad i = 1,...,c.$$

3. Moreover, the test statistic is

$$Q^2 = Q^2(\mathbf{Y}) = \sum_{i=1}^{c} \{n_i \hat{\mathbf{U}}_i' \hat{\mathbf{B}}^{-1} \hat{\mathbf{U}}_i\},$$

   where $\hat{\mathbf{B}} = \mathbf{AVE}\left\{\hat{\mathbf{U}}_{ij}\hat{\mathbf{U}}_{ij}'\right\}$
4. Under the null hypothesis $H_0: \; \mu_1 = \cdots = \mu_c$ and under some weak assumptions, the limiting distribution of $Q^2(\mathbf{Y})$ is $\chi^2_{(c-1)p}$.

The distributional assumptions needed for the asymptotical results are now the following.

Assumption 3 *The density function of* $\mathbf{y}_{ij}$ *is bounded and continuous. Moreover, the spatial median of* $\mathbf{y}_{ij} - \boldsymbol{\mu}_i$ *is unique and zero.*

The assumptions are weaker than in the case of the identity score in the sense that no moment assumptions are needed. As in the identity score case, we need assumptions on the limiting behavior of the sample sizes.

Assumption 4

$$\frac{n_i}{n} \to \lambda_i, \quad i = 1,...,c, \quad \textit{as } n \to \infty$$

with $0 < \lambda_i < 1$, $i = 1,...,c$.

Then we can prove the next theorem.

Theorem 11.2.

- *Under Assumptions 3 and 4 and under the null hypothesis* $H_0 : \boldsymbol{\mu}_1 = \cdots = \boldsymbol{\mu}_c$, *the limiting distribution of* Q^2 *is a chi-square distribution with* $(c-1)p$ *degrees of freedom.*
- *Under Assumptions 3 and 4 and under the sequence of alternatives* $H_n : \boldsymbol{\mu}_i = n^{-1/2}\boldsymbol{\delta}_i$, $i = 1,...,c$, *where* $\sum_{i=1}^{c} \lambda_i \boldsymbol{\delta}_i = 0$, *the limiting distribution of the test statistic* Q^2 *is a chi-square distribution with* $(c-1)p$ *degrees of freedom and the noncentrality parameter*

$$\sum_{i=1}^{c} \lambda_i \boldsymbol{\delta}_i' \mathbf{A}\mathbf{B}^{-1}\mathbf{A}\boldsymbol{\delta}_i,$$

where, as in the one-sample case,

$$\mathbf{A} = \mathbf{E}\left\{ \frac{1}{|\mathbf{y}_{ij} - \boldsymbol{\mu}_i|} \left[\mathbf{I}_p - \frac{(\mathbf{y}_{ij} - \boldsymbol{\mu}_i)(\mathbf{y}_{ij} - \boldsymbol{\mu}_i)'}{|\mathbf{y}_{ij} - \boldsymbol{\mu}_i|^2} \right] \right\}$$

and

$$\mathbf{B} = \mathbf{E}\left\{ \frac{(\mathbf{y}_{ij} - \boldsymbol{\mu}_i)(\mathbf{y}_{ij} - \boldsymbol{\mu}_i)'}{|\mathbf{y}_{ij} - \boldsymbol{\mu}_i|^2} \right\}.$$

Proof. We first assume that the null hypothesis is true with $\boldsymbol{\mu}_1 = \cdots = \boldsymbol{\mu}_c = \mathbf{0}$. Then $E(\mathbf{U}(\mathbf{y}_{ij})) = \mathbf{0}$. Write

$$\mathbf{U}_{ij} = \mathbf{U}(\mathbf{y}_{ij}), \quad i = 1,...,c; \; j = 1,...,n_i,$$

for the "theoretical scores" and

$$\bar{\mathbf{U}}_i = \mathbf{AVE}_j \{\mathbf{U}_{ij}\} \quad \text{and} \quad \bar{\mathbf{U}} = \mathbf{AVE}\{\mathbf{U}_{ij}\}.$$

As the null hypothesis is true and we are back in the one-sample case and

$$\sqrt{n}\hat{\boldsymbol{\mu}} = \sqrt{n}\mathbf{A}^{-1}\bar{\mathbf{U}} + o_P(1).$$

Recall the results in Section 6.2. Using the results in Section 6.1.1, we can conclude that

$$\sqrt{n_i}\hat{\mathbf{U}}_i = \sqrt{n_i}\bar{\mathbf{U}}_i + \sqrt{n_i}\mathbf{A}\hat{\mu} + o_P(1) = \sqrt{n_i}\left(\bar{\mathbf{U}}_i - \bar{\mathbf{U}}\right) + o_P(1).$$

But then we are back in the regular MANOVA case with the identity score function and the $\mathbf{U}_{ij}$ as observation vectors, and the null result follows from Theorem 11.1.

Consider next a sequence of alternatives

$$H_n: \ \mu_i = \mu + \frac{1}{\sqrt{n}}\delta_i, \quad i = 1, ..., c,$$

where $\sum_{i=1}^{c} \lambda_i \delta_i = 0$. Write

$$\tilde{\delta} = (\sqrt{\lambda_1}\delta'_1, ..., \sqrt{\lambda_{c-1}}\delta'_{c-1})'.$$

Then, under the alternative sequence, the limiting distribution of

$$(\sqrt{n_1}\hat{\mathbf{U}}'_1, ..., \sqrt{n_{c-1}}\hat{\mathbf{U}}'_{c-1})'$$

is multivariate normal with mean vector

$$(\mathbf{I}_{c-1} \otimes \mathbf{A})\tilde{\delta}$$

and covariance matrix

$$\left(\mathbf{I}_{c-1} - \mathbf{dd}'\right) \otimes \mathbf{B},$$

where $\mathbf{d} = (\sqrt{\lambda_1}, ..., \sqrt{\lambda_{c-1}})'$. Then, under the alternative sequence, the limiting distribution of Q^2 is the noncentral chi-square distribution with $(c-1)p$ degrees of freedom and noncentrality parameter

$$\tilde{\delta}'\left(\left(\mathbf{I}_{c-1} + \frac{\mathbf{dd}'}{1 - \mathbf{d}'\mathbf{d}}\right) \otimes \mathbf{AB}^{-1}\mathbf{A}\right)\tilde{\delta} = \sum_{i=1}^{c} \lambda_i \delta'_i \mathbf{AB}^{-1}\mathbf{A}\delta_i.$$

Again, if Δ is the $n \times p$ matrix of the location centers of the observations under the alternative and $\hat{\Delta} = (\mathbf{I}_n - \mathbf{P}_{\mathbf{1}_n})\Delta$, one can use Theorem 11.2 to approximate the distribution of Q^2. It has an approximate noncentral chi-square distribution with $(c-1)p$ degrees of freedom and noncentrality parameter

$$tr(\hat{\Delta}'\mathbf{P_X}\hat{\Delta}\mathbf{AB}^{-1}\mathbf{A}).$$

Totally nonparametric model. The permutation version of the MANOVA test based on the spatial sign score is obtained as follows. As before, write the dataset as

$$(\mathbf{X}, \mathbf{Y}),$$

where the ith column of the matrix $\mathbf{X}$ indicates the membership of the ith sample. Then transform the matrix $\mathbf{Y}$ to the inner centered score matrix $\hat{\mathbf{U}}$. Then again

$$Q^2(\mathbf{Y}) = n \cdot tr\left(\hat{\mathbf{U}}'\mathbf{P_X}\hat{\mathbf{U}}(\hat{\mathbf{U}}'\hat{\mathbf{U}})^{-1}\right)$$

and

$$Q^2(\mathbf{PY}) = n \cdot tr\left(\hat{\mathbf{U}}'\mathbf{PP_X P}\hat{\mathbf{U}}(\hat{\mathbf{U}}'\hat{\mathbf{U}})^{-1}\right).$$

As before, the exact p value is given by

$$\mathbf{E_P}\left(Q^2(\mathbf{PY}) \geq Q^2(\mathbf{Y})\right),$$

where $\mathbf{P}$ is uniformly distributed over the set of all $n!$ different permutation matrices.

Affine invariant test. Unfortunately, the spatial sign test statistic discussed so far is not affine invariant. An affine invariant version of the test is obtained if one uses inner centering and inner standardization:

Test statistic using inner centering and inner standardization.

1. Find a shift vector $\hat{\mu}$ and full rank transformation matrix $\mathbf{S}^{-1/2}$ such that scores for
$$\hat{\mathbf{U}}_{ij} = \mathbf{U}(\mathbf{S}^{-1/2}(\mathbf{y}_{ij} - \hat{\mu})), \quad i = 1, \ldots, c; \; j = 1, \ldots, n_i,$$
are standardized in the sense that
$$\mathbf{AVE}\left\{\hat{\mathbf{U}}_{ij}\right\} = \mathbf{0} \quad \text{and} \quad p \cdot \mathbf{AVE}\left\{\hat{\mathbf{U}}_{ij}\hat{\mathbf{U}}_{ij}'\right\} = \mathbf{I}_p.$$
2. The test statistic for testing whether the ith sample differs from the others is then based on
$$\hat{\mathbf{U}}_i = \mathbf{AVE}_j\left\{\hat{\mathbf{U}}_{ij}\right\}, \quad i = 1, \ldots, n.$$
3. The several-sample location test statistic
$$Q^2 = p \cdot \sum n_i |\hat{\mathbf{U}}_i|^2.$$
4. Under the null hypothesis $H_0: \mu_1 = \cdots = \mu_c$ and under some weak assumptions, the limiting distribution of Q^2 is $\chi^2_{(c-1)p}$.

Note that if one transforms $\mathbf{Y} \to \hat{\mathbf{U}}$ where the scores in $\hat{\mathbf{U}}$ are inner centered and inner standardized then simply

$$Q^2(\mathbf{Y}) = p \cdot |\mathbf{P_X}\hat{\mathbf{U}}|^2.$$

Now

$$Q^2(\mathbf{PY}) = p \cdot |\mathbf{P_X P}\hat{\mathbf{U}}|^2$$

and the p values of the exact test version are easily calculated.

11.4 The tests based on spatial ranks

This approach uses the spatial rank function $\mathbf{R}(\mathbf{y}) = \mathbf{R}_{\mathbf{Y}}(\mathbf{y})$. For a dataset $\mathbf{Y} = (\mathbf{y}_1, ..., \mathbf{y}_n)'$, the spatial (centered) rank function is

$$\mathbf{R}(\mathbf{y}) = \mathbf{AVE}\{\mathbf{U}(\mathbf{y} - \mathbf{y}_i)\}.$$

Remember that the spatial ranks

$$\mathbf{R}_i = \mathbf{R}(\mathbf{y}_i), \quad i = 1, ..., n,$$

are automatically centered; that is, $\sum_{i=1}^{n} \mathbf{R}_i = \mathbf{0}$, and no inner centering is therefore needed. The tests extend the two-sample *Wilcoxon-Mann-Whitney test* and the several-sample *Kruskal-Wallis test*.

We assume that the observations $\mathbf{y}_{ij}$ are generated by

$$\mathbf{y}_{ij} = \mu_i + \Omega \varepsilon_{ij}, \quad i = 1, ..., c; \; j = 1, ..., n_i,$$

where ε_{ij} are independent, centered, and standardized observations (in the sense described later) with a joint cumulative distribution function F. Again, we wish to test the null hypothesis

$$H_0: \; \mu_1 = \cdots = \mu_c.$$

Then the first testing procedure is the following.

Test statistic with outer standardization.

1. Centered spatial rank scores are

$$\mathbf{R}_{ij} = \mathbf{R}_{\mathbf{Y}}(\mathbf{y}_{ij}), \quad i = 1, ..., c; \; j = 1, ..., n_i,$$

   which satisfy $\mathbf{AVE}\{\mathbf{R}_{ij}\} = \mathbf{0}$.
2. The test statistic for testing whether the ith sample differs from the others is based on

$$\bar{\mathbf{R}}_i = \mathbf{AVE}_j\{\mathbf{R}_{ij}\}, \quad i = 1, ..., c.$$

3. Moreover, the test statistic for $H_0: \; \mu_1 = \cdots = \mu_c$ is

$$Q^2 = Q^2(\mathbf{Y}) = \sum_{i=1}^{c} \{n_i \bar{\mathbf{R}}_i' \hat{\mathbf{B}}^{-1} \bar{\mathbf{R}}_i\},$$

   where $\hat{\mathbf{B}} = \mathbf{RCOV}(\mathbf{Y}) = \mathbf{AVE}\left\{\mathbf{R}_{ij}\mathbf{R}_{ij}'\right\}$.
4. Under the null hypothesis and under some weak assumptions, the limiting distribution of $Q^2(\mathbf{Y})$ is $\chi^2_{(c-1)p}$.

The distributional assumptions needed for the asymptotical results are now as follows.

Assumption 5 *The density function of* $\mathbf{y}_{ij}$ *is bounded and continuous. Moreover, the spatial median of* $\mathbf{y}_{ij} - \mathbf{y}_{rs}$ *is unique* $(\mu_i - \mu_r)$.

Again, no moment assumptions are needed, and the limiting behavior of the sample sizes is as stated in Assumption 4.

Then we can prove the next theorem.

Theorem 11.3.

- *Under Assumptions 4 and 5 and under the null hypothesis* $H_0: \mu_1 = \cdots = \mu_c$, *the limiting distribution of* Q^2 *is a chi square distribution with* $(c-1)p$ *degrees of freedom.*
- *Under Assumptions 4 and 5 and under the sequence of alternatives* $H_n: \mu_i = n^{-1/2}\delta_i$, $i = 1,...,c$, *where* $\sum_{i=1}^c \lambda_i \delta_i = 0$, *the limiting distribution of the test statistic* Q^2 *is a chi-square distribution with* $(c-1)p$ *degrees of freedom and a noncentrality parameter*

$$\sum_{i=1}^{c} \lambda_i \delta_i' \mathbf{A}\mathbf{B}^{-1}\mathbf{A}\delta_i,$$

where now, with expected values calculated in the null case,

$$\mathbf{A} = \mathbf{E}\left\{\frac{1}{|\mathbf{y}_1 - \mathbf{y}_2|}\left[\mathbf{I}_p - \frac{(\mathbf{y}_1 - \mathbf{y}_2)(\mathbf{y}_1 - \mathbf{y}_2)'}{|\mathbf{y}_1 - \mathbf{y}_2|^2}\right]\right\}$$

and

$$\mathbf{B} = \mathbf{E}\left\{\frac{(\mathbf{y}_1 - \mathbf{y}_2)(\mathbf{y}_1 - \mathbf{y}_3)'}{|\mathbf{y}_1 - \mathbf{y}_2| \cdot |\mathbf{y}_1 - \mathbf{y}_3|}\right\}.$$

Proof. We first assume that the null hypothesis is true; that is, $E(\mathbf{U}(\mathbf{y}_{ij} - \mathbf{y}_{rs})) = \mathbf{0}$ for all i, j, r, s. The test statistic

$$\bar{\mathbf{R}}_i = \mathbf{AVE}_j\{\mathbf{R}_{ij}\},$$

is a two-sample U-statistic and, under the null hypothesis, asymptotically equivalent with its projection statistic

$$\mathbf{AVE}_j\{\mathbf{R}_F(\mathbf{y}_{ij})\} - \mathbf{AVE}_{ij}\{\mathbf{R}_F(\mathbf{y}_{ij})\},$$

where $\mathbf{R}_F(\mathbf{y}) = \mathbf{E}_F(\mathbf{U}(\mathbf{y} - \mathbf{y}_{ij}))$ is the theoretical (population) centered rank function. But now we are back in the regular MANOVA case with identity score function and the $\mathbf{R}_F(\mathbf{y}_{ij})$ as observed vectors, and the null result follows from Theorem 11.1.

Next consider a sequence of alternatives

$$H_n: \ \mu_i = \mu + \frac{1}{\sqrt{n}}\delta_i, \quad i = 1,...,c,$$

where $\sum_{i=1}^c \lambda_i \delta_i = 0$. Write

$$\tilde{\delta} = (\sqrt{\lambda_1}\delta_1', ..., \sqrt{\lambda_{c-1}}\delta_{c-1}')'.$$

Then, under the alternative sequence, the limiting distribution of

$$(\sqrt{n_1}\bar{\mathbf{R}}_1', ..., \sqrt{n_{c-1}}\bar{\mathbf{R}}_{c-1}')'$$

is a multivariate normal with the mean vector

$$(\mathbf{I}_{c-1} \otimes \mathbf{A})\tilde{\delta}$$

and the covariance matrix

$$(\mathbf{I}_{c-1} - \mathbf{d}\mathbf{d}') \otimes \mathbf{B},$$

where $\mathbf{d} = (\sqrt{\lambda_1}, ..., \sqrt{\lambda_{c-1}})'$. Then, under the alternative sequence, the limiting distribution of Q^2 is a noncentral chi-square distribution with $(c-1)p$ degrees of freedom and the noncentrality parameter

$$\tilde{\delta}'\left(\left(\mathbf{I}_{c-1} + \frac{\mathbf{d}\mathbf{d}'}{1-\mathbf{d}'\mathbf{d}}\right) \otimes \mathbf{A}\mathbf{B}^{-1}\mathbf{A}\right)\tilde{\delta} = \sum_{i=1}^{c} \lambda_i \delta_i' \mathbf{A}\mathbf{B}^{-1}\mathbf{A}\delta_i.$$

Totally nonparametric model. The permutation version of the MANOVA test based on the spatial ranks is obtained in a similar way as in the case of other scores. As before an $n \times c$ matrix $\mathbf{X}$ identifies the sample so that the ith column of the matrix $\mathbf{X}$ indicates the membership in the ith sample. The $n \times p$ data matrix $\mathbf{Y}$ is transformed to the matrix $\mathbf{R}$ of centered spatial signs. Then

$$Q^2(\mathbf{Y}) = n \cdot tr\left(\mathbf{R}'\mathbf{P}_{\mathbf{X}}\mathbf{R}(\mathbf{R}'\mathbf{R})^{-1}\right)$$

and

$$Q^2(\mathbf{P}\mathbf{Y}) = n \cdot tr\left(\mathbf{R}'\mathbf{P}\mathbf{P}_{\mathbf{X}}\mathbf{P}\mathbf{R}(\mathbf{R}'\mathbf{R})^{-1}\right)$$

and the exact p-values are obtained as before.

The test statistic is not affine invariant, however. An affine invariant modification of the test is obtained if one uses inner centering and inner standardization as follows.

Test statistic using inner standardization.

1. Find a full rank transformation matrix $\mathbf{S}^{-1/2}$ such that

$$\mathbf{RCOV}(\mathbf{Y}\mathbf{S}^{-1/2}) \propto \mathbf{I}_p.$$

   Write

$$\hat{\mathbf{R}}_{ij} = \mathbf{R}_{\mathbf{Y}\mathbf{S}^{-1/2}}(\mathbf{S}^{-1/2}\mathbf{y}_{ij}), \quad i = 1, ..., c;\ j = 1, ..., n_i.$$

2. The test statistic for testing whether the ith sample differs from the others is then based on

$$\tilde{\mathbf{R}}_i = \mathbf{AVE}_j \left\{ \hat{\mathbf{R}}_{ij} \right\}, \quad i = 1, ..., n.$$

3. The several-sample location test statistic is then

$$Q^2 = np \cdot \frac{\sum n_i |\tilde{\mathbf{R}}_i|^2}{\sum_i \sum_j |\hat{\mathbf{R}}_{ij}|^2}.$$

4. Under the null hypothesis and under some weak assumptions, the limiting distribution of $Q^2(\mathbf{Y})$ is $\chi^2_{(c-1)p}$.

11.5 Estimation of the treatment effects

Let the data matrix

$$\mathbf{Y} = (\mathbf{Y}'_1, ..., \mathbf{Y}'_c)'$$

consist of c independent random samples

$$\mathbf{Y}_i = (\mathbf{y}_{i1}, ..., \mathbf{y}_{in_i})', \quad i = 1, ..., c,$$

from p-variate distributions. Write $n = n_1 + \cdots + n_c$ and, for the asymptotic results, assume that

$$n \to \infty \quad \text{and} \quad \frac{n_i}{n} \to \lambda_i, \quad i = 1, ..., c,$$

where $0 < \lambda_i < 1$, $i = 1, ..., c$. We again assume that the independent p-variate observation vectors $\mathbf{y}_{ij}$, are generated by

$$\mathbf{y}_{ij} \; = \; \mu_i + \Omega \, \varepsilon_{ij}, \quad i = 1, ..., c; \; j = 1, ..., n_i,$$

where the ε_{ij} are the standardized and centered vectors all having the same unknown distribution. As before, $\mu_1, ..., \mu_c$ are unknown location centers and the matrix $\Sigma = \Omega\Omega' > 0$ is a joint unknown scatter matrix.

We now wish to estimate unknown differences $\Delta_{ij} = \mu_j - \mu_i$, $i, j = 1, ..., c$. The three scores (identity, spatial sign, and spatial rank) yield the following estimates,

- The difference between the sample mean vectors,
- The difference between the sample spatial medians, and
- The two-sample Hodges-Lehmann estimate,

correspondingly. For symmetrical distributions, all three estimates estimate the same population quantity Δ_{ij}. We now have a closer look at these three estimates.

The difference between the sample means. The first estimate of Δ_{ij} is

$$\hat{\Delta}_{ij} = \bar{\mathbf{y}}_j - \bar{\mathbf{y}}_i, \quad i, j = 1, ..., c.$$

If $\Sigma = \mathbf{COV}(\mathbf{y}_{ij})$ exists then, using the central limit theorem,

$$\sqrt{n}(\hat{\Delta}_{ij}-\Delta_{ij}) \to_d N_p\left(\mathbf{0}, \left(\frac{1}{\lambda_1}+\frac{1}{\lambda_2}\right)\Sigma\right), \quad i \neq j.$$

Of course, Σ is unknown but can be estimated with

$$\hat{\Sigma} = \mathbf{AVE}\{(\mathbf{y}_{ij}-\bar{\mathbf{y}}_i)(\mathbf{y}_{ij}-\bar{\mathbf{y}}_i)'\}.$$

Naturally the estimates are affine equivariant and satisfy

$$\hat{\Delta}_{ij} = \hat{\Delta}_{ik} + \hat{\Delta}_{kj}, \quad \text{for all } i,k,j = 1,...,c.$$

The difference between the sample spatial medians. Let now $\hat{\mu}_i$ be the spatial median calculated for $\mathbf{Y}_i$, $i = 1,...,c$. Our second estimate for Δ_{ij} is then

$$\hat{\Delta}_{ij} = \hat{\mu}_j - \hat{\mu}_i, \quad i,j = 1,...,c.$$

If

$$\mathbf{A}(\mathbf{y}) = \frac{1}{|\mathbf{y}|}\left(\mathbf{I}_p - \frac{\mathbf{y}\mathbf{y}'}{|\mathbf{y}|}\right) \quad \text{and} \quad \mathbf{B}(\mathbf{y}) = \frac{\mathbf{y}\mathbf{y}'}{|\mathbf{y}|^2}$$

and

$$\mathbf{A} = \mathbf{E}\{\mathbf{A}(\mathbf{y}_{ij}-\mu_i)\} \text{ and } \mathbf{B} = \mathbf{E}\{\mathbf{B}(\mathbf{y}_{ij}-\mu_i)\}$$

then, using Theorem 7.4,

$$\sqrt{n}(\hat{\Delta}_{ij}-\Delta_{ij}) \to_d N_p\left(\mathbf{0}, \left(\frac{1}{\lambda_1}+\frac{1}{\lambda_2}\right)\mathbf{A}^{-1}\mathbf{B}\mathbf{A}^{-1}\right).$$

No moment assumptions are needed here. Of course, $\mathbf{A}$ and $\mathbf{B}$ are unknown but can be estimated by

$$\hat{\mathbf{A}} = \mathbf{AVE}\left[\mathbf{A}(\mathbf{y}_{ij}-\hat{\mu}_i)\right] \quad \text{and} \quad \hat{\mathbf{B}} = \mathbf{AVE}\left[\mathbf{B}(\mathbf{y}_{ij}-\hat{\mu}_i)\right].$$

The estimates satisfy

$$\hat{\Delta}_{ij} = \hat{\Delta}_{ik} + \hat{\Delta}_{kj}, \quad \text{for all } i,k,j = 1,...,c,$$

but they are not affine equivariant.

An affine equivariant estimator is obtained with the inner standardization as follows. Let μ_i, $i = 1,...,c$, be p-vectors and $\mathbf{S} > 0$ a symmetric $p \times p$ matrix, and define

$$\varepsilon_{ij} = \varepsilon_{ij}(\mu_i, \mathbf{S}) = \mathbf{S}^{-1/2}(\mathbf{y}_{ij}-\mu_i), \quad i = 1,...,c; \; j = 1,...,n_i.$$

The Hettmansperger-Randles (HR) estimates of the location centers and the joint scatter matrix are the values of $\mu_1,...,\mu_c$ and $\mathbf{S}$ that simultaneously satisfy

$$\mathbf{AVE}_j\left\{\mathbf{U}(\varepsilon_{ij})\right\} = \mathbf{0}, \quad i = 1,...,c,$$

and

$$p \cdot \mathbf{AVE}\{\mathbf{U}(\varepsilon_{ij})\mathbf{U}(\varepsilon_{ij})'\} = \mathbf{I}_p.$$

The two-sample Hodges-Lehmann estimate. The two-sample Hodges-Lehmann estimate $\hat{\Delta}_{ij}$ is the spatial median calculated from the pairwise differences

$$\mathbf{y}_{jr} - \mathbf{y}_{is}, \quad r = 1, ..., n_j; \ s = 1, ..., n_i.$$

If now

$$\mathbf{A}(\mathbf{y}) = \frac{1}{|\mathbf{y}|}\left(\mathbf{I} - \frac{\mathbf{y}\mathbf{y}'}{|\mathbf{y}|}\right) \quad \text{and} \quad \mathbf{B}(\mathbf{y}_1, \mathbf{y}_2) = \frac{\mathbf{y}_1\mathbf{y}_2'}{|\mathbf{y}_1| \cdot |\mathbf{y}_2|}$$

and

$$\mathbf{A} = \mathbf{E}\{\mathbf{A}(\mathbf{y}_{11} - \mathbf{y}_{12})\} \quad \text{and} \quad \mathbf{B} = \mathbf{E}\{\mathbf{B}(\mathbf{y}_{11} - \mathbf{y}_{12}, \mathbf{y}_{11} - \mathbf{y}_{13})\},$$

then again

$$\sqrt{n}(\hat{\Delta}_{ij} - \Delta_{ij}) \to_d N_p\left(\mathbf{0}, \left(\frac{1}{\lambda_1} + \frac{1}{\lambda_2}\right)\mathbf{A}^{-1}\mathbf{B}\mathbf{A}^{-1}\right).$$

Again, $\mathbf{A}$ and $\mathbf{B}$ are unknown but can be estimated by

$$\hat{\mathbf{A}} = \mathbf{AVE}\{\mathbf{A}(\mathbf{y}_{jr} - \mathbf{y}_{is} - \hat{\Delta}_{ij})\}$$

and by

$$\hat{\mathbf{B}} = \mathbf{AVE}\{\mathbf{B}(\mathbf{y}_{jr} - \mathbf{y}_{is} - \hat{\Delta}_{ij}, \mathbf{y}_{jr} - \mathbf{y}_{kl} - \hat{\Delta}_{kj})\},$$

respectively.

Unfortunately, the estimates $\hat{\Delta}_{ij}$ are not any more compatible in the sense that $\hat{\Delta}_{ij} = \hat{\Delta}_{ik} + \hat{\Delta}_{kj}$. To overcome this problem, one can first, using the kth sample as a reference sample, find an estimate for the difference between ith and jth sample as

$$\tilde{\Delta}_{ij \cdot k} = \hat{\Delta}_{ik} + \hat{\Delta}_{kj}, \quad k = 1, ..., c,$$

and then take the weighted average, Spjøtvoll's estimator (Spjøtvoll (1968))

$$\tilde{\Delta}_{ij} = \frac{1}{n}\sum_{k=1}^{c} n_k \tilde{\Delta}_{ij \cdot k}.$$

Then

$$\tilde{\Delta}_{ij} = \tilde{\Delta}_{ik} + \tilde{\Delta}_{kj}, \quad \text{for all } i, k, j = 1, ..., c.$$

One can easily show that

$$\sqrt{n}(\tilde{\Delta}_{ij} - \hat{\Delta}_{ij}) \to_P \mathbf{0}$$

so that the limiting distributions of $\sqrt{n}(\tilde{\Delta}_{ij} - \Delta_{ij})$ and $\sqrt{n}(\hat{\Delta}_{ij} - \Delta_{ij})$ are the same. See Nevalainen et al. (2007c).

An affine equivariant estimator is obtained with the transformation and retransformation techique as follows. Let $\hat{\Delta}_{ij}$, $i,j=1,...,c$, be p-vectors and $\mathbf{S}>0$ a symmetric $p\times p$ that simultaneously satisfy

$$\mathbf{AVE}_{rs}\left\{\mathbf{U}\left(\mathbf{S}^{-1/2}(\mathbf{y}_{jr}-\mathbf{y}_{is}-\hat{\Delta}_{ij})\right)\right\}=\mathbf{0},\quad \text{for all } i,j=1,...,c$$

and

$$\mathbf{AVE}\left\{\mathbf{B}\left(\mathbf{S}^{-1/2}(\mathbf{y}_{jr}-\mathbf{y}_{is}-\hat{\Delta}_{ij}),\mathbf{S}^{-1/2}(\mathbf{y}_{jr}-\mathbf{y}_{kl}-\hat{\Delta}_{kj})\right)\right\}\propto\mathbf{I}_p.$$

11.6 An example: Egyptian skulls from three epochs.

As an example, we consider the classical dataset on Egyptian skulls from three different epochs. The same data were analyzed in Johnson and Wichern (1998). The three epochs were time periods in years around 4000, around 3300, and around 1850 BC. Thirty skulls were measured for each time period, and the four measured variables denoted by mb (maximal breadth), bh (basibregmatic height), bl (basialveolar length), and nh (nasal height). We wish to consider if there are any differences in the mean skull sizes among the time periods. See below the first description of the dataset. The observations are plotted in Figure 11.1.

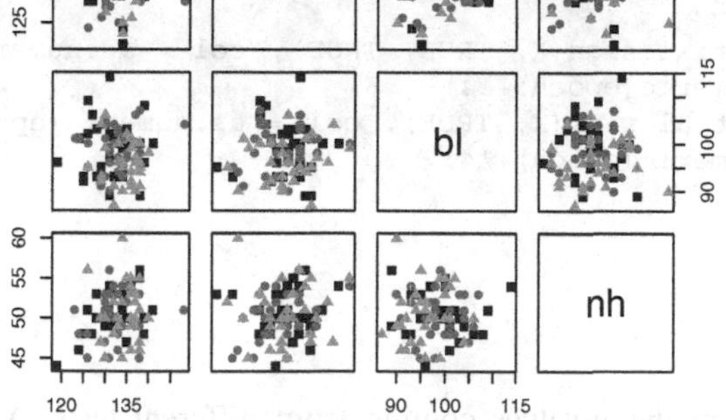

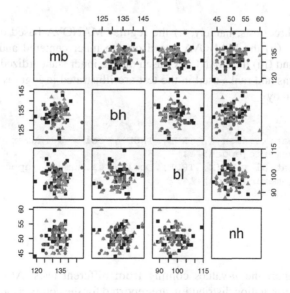

Fig. 11.1 Egyptian skulls data.

```
> library(MNM)
> library(HSAUR)
>
> # using skulls as in Johnson and Wichern
>
> SKULLS <- skulls[1:90,]
> levels(SKULLS epoch)<- c(levels(SKULLS epoch)[1:3], NA, NA)
> summary(SKULLS)
     epoch              mb                bh
 c4000BC:30    Min.    :119    Min.    :121
 c3300BC:30    1st Qu.:130    1st Qu.:130
 c1850BC:30    Median :133    Median :134
               Mean    :133    Mean    :133
               3rd Qu.:136    3rd Qu.:136
               Max.    :148    Max.    :145
                    bl                nh
               Min.    : 87.0    Min.    :44.0
               1st Qu.: 94.2    1st Qu.:48.0
               Median : 98.0    Median :50.0
               Mean    : 98.1    Mean    :50.4
               3rd Qu.:101.0    3rd Qu.:53.0
               Max.    :114.0    Max.    :60.0
> X <- SKULLS[,2:5]
> epoch <- SKULLS epoch
> pairs(X, col = as.numeric(epoch), pch=as.numeric(epoch)+14)
```

We compare three tests, namely, (i) the regular MANOVA based on the identity score function, (ii) the MANOVA based on the inner centered and standardized spatial signs, and (iii) the MANOVA based on the inner standardized spatial ranks. The spatial signs and ranks used in the test are illustrated in Figures 11.2 and 11.3 which are given by

```
> pairs(spatial.sign(X, TRUE, TRUE), col = as.numeric(epoch),
  pch=as.numeric(epoch)+14)
> pairs(spatial.rank(X, TRUE), col = as.numeric(epoch),
  pch=as.numeric(epoch)+14)
```

We next report the p-values coming from different tests. Also the p-values based on the permutation distribution are reported for the comparison. The classical MANOVA test produces

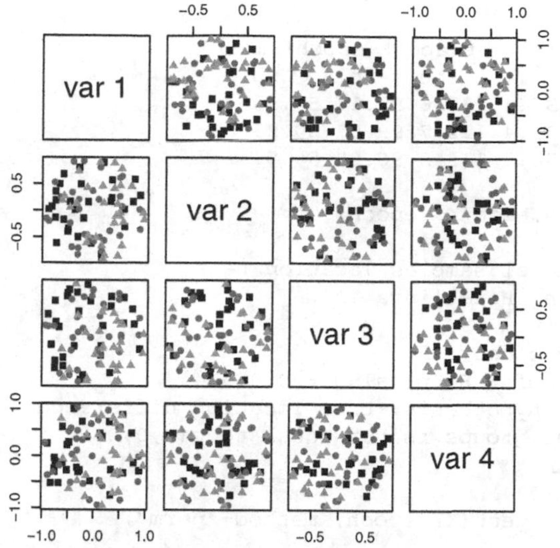

Fig. 11.2 Egyptian skulls data. Inner centered and inner standardized spatial signs.

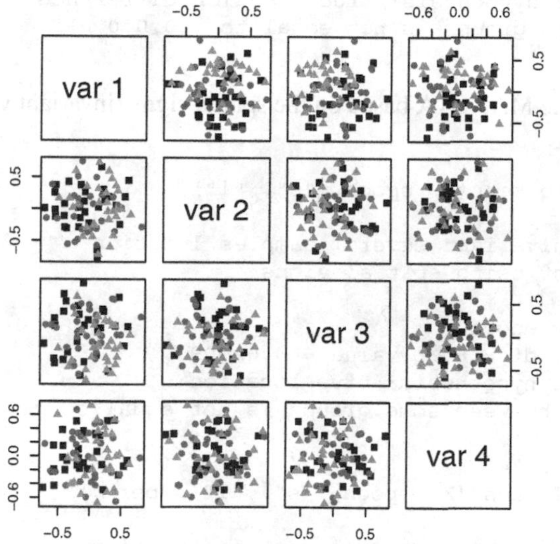

Fig. 11.3 Egyptian skulls data. Inner standardized spatial ranks.

```
>
> aggregate(X,list(epoch=epoch),mean)
    epoch    mb    bh    bl    nh
1 c4000BC 131.4 133.6 99.17 50.53
2 c3300BC 132.4 132.7 99.07 50.23
3 c1850BC 134.5 133.8 96.03 50.57
>
> mv.Csample.test(X, epoch)

        Several samples location test
        using Hotellings T2

data:  X by epoch
Q.2 = 15.5, df = 8, p-value = 0.05014
alternative hypothesis: true location difference
between some groups is not equal to c(0,0,0,0)

> set.seed(1234)
> mv.Csample.test(X, epoch, method="perm")

        Several samples location test
        using Hotellings T2

data:  X by epoch
Q.2 = 15.5, replications = 1000, p-value = 0.05
alternative hypothesis: true location difference
between some groups is not equal to c(0,0,0,0)
```

If one uses the MANOVA based on the spatial signs (invariant version), one gets

```
> mv.Csample.test(X, epoch, "s", "i")

        Equivariant several samples location
        test using spatial signs

data:  X by epoch
Q.2 = 17.1, df = 8, p-value = 0.02909
alternative hypothesis: true location
difference between some groups is not equal to c(0,0,0,0)

> set.seed(1234)
> mv.Csample.test(X, epoch, "s", "i", "perm")

        Equivariant several samples location
        test using spatial signs

data:  X by epoch
Q.2 = 17.1, replications = 1000, p-value = 0.032
alternative hypothesis: true location difference
between some groups is not equal to c(0,0,0,0)
```

With the invariant the rank-based MANOVA one gets the following results.

```
> mv.Csample.test(X, epoch, "r", "i")

        Equivariant several samples location
        test using spatial ranks

data:  X by epoch
Q.2 = 16.67, df = 8, p-value = 0.03372
alternative hypothesis: true location difference
between some groups is not equal to c(0,0,0,0)

> set.seed(1234)
> mv.Csample.test(X, epoch, "r", "i", "perm")

        Equivariant several samples location
        test using spatial ranks

data:  X by epoch
Q.2 = 16.67, replications = 1000, p-value = 0.034
alternative hypothesis: true location difference
between some groups is not equal to c(0,0,0,0)
```

We end this example with the comparison of different two-sample location estimates. The estimates are (i) the difference of the mean vectors, (ii) the difference of the affine equivariant spatial medians, and (iii) the affine equivariant two-sample Hodges-Lehmann estimate. We compare the first and the last epoch, and get the following.

```
> SKULLS.13 <- SKULLS[c(1:30,61:90),]
> levels(SKULLS.13 epoch)<-
  c(levels(SKULLS.13 epoch)[1], NA,
  levels(SKULLS.13 epoch)[3])
> X.13 <- SKULLS.13[,2:5]
> epoch.13 <- SKULLS.13 epoch
> EST1.13 <- mv.2sample.est(X.13,epoch.13)
> summary(EST1.13)
The difference between sample mean vectors
of X.13 by epoch.13 is:
     mb      bh      bl      nh
-3.1000 -0.2000  3.1333 -0.0333

And has the covariance matrix:
        mb     bh      bl      nh
mb  1.2810 0.1646 -0.0107  0.2715
bh  0.1646 1.4920  0.0933  0.0101
```

```
bl -0.0107 0.0933  1.8450 -0.0083
nh  0.2715 0.0101 -0.0083  0.6745
> EST3.13 <- mv.2sample.est(X.13,epoch.13,"s" ,"i")
> summary(EST3.13)
The difference between equivariant spatial medians
of X.13 by epoch.13 is:
[1] -3.6719 -0.1304  3.8302  0.1285

And has the covariance matrix:
        [,1]    [,2]    [,3]   [,4]
[1,]  1.3423  0.4713 -0.1570 0.2580
[2,]  0.4713  1.4661 -0.0794 0.2313
[3,] -0.1570 -0.0794  1.7255 0.0738
[4,]  0.2580  0.2313  0.0738 0.5072
> EST5.13 <- mv.2sample.est(X.13,epoch.13,"r" ,"i")
> summary(EST5.13)
The equivariant spatial Hodges-Lehmann estimator for
location difference of X.13 by epoch.13 is:
        [,1]    [,2]    [,3]    [,4]
[1,] -3.3465 -0.1283  3.3335  0.1556

And has the covariance matrix:
        [,1]    [,2]    [,3]   [,4]
[1,]  1.2440  0.2697 -0.0018 0.2701
[2,]  0.2697  1.3138 -0.0741 0.1081
[3,] -0.0018 -0.0741  1.7973 0.0605
[4,]  0.2701  0.1081  0.0605 0.5920
>
> plotMvloc(EST1.13, EST3.13, EST5.13, lty.ell = 1:3,
pch.ell = 14:17)
```

The differences in the estimates also seem to be minimal, as can be seen in Figure 11.4. The assumption on the multivariate normality of the data may be realistic here. If the last observation is changed to $(200,200,200,200)'$ to be an outlier then the estimate and the confidence ellipsoid based on the regular mean vectors also changes dramatically as can be seen in Figure 11.5.

11.7 References and other approaches

See Möttönen and Oja (1995), Choi and Marden (1997), Marden (1999a), Visuri et al. (2003), Oja and Randles (2004), and Nevalainen et al. (2007c) for different uses of spatial signs and ranks in the multivariate several-sample location problem. Mardia (1967) considered the bivariate problems. See Nevalainen and Oja (2006)

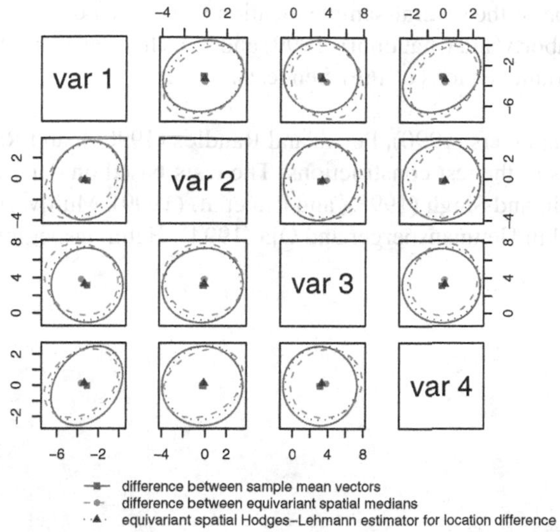

Fig. 11.4 Egyptian skulls data. Estimates of the difference between the first and last epochs.

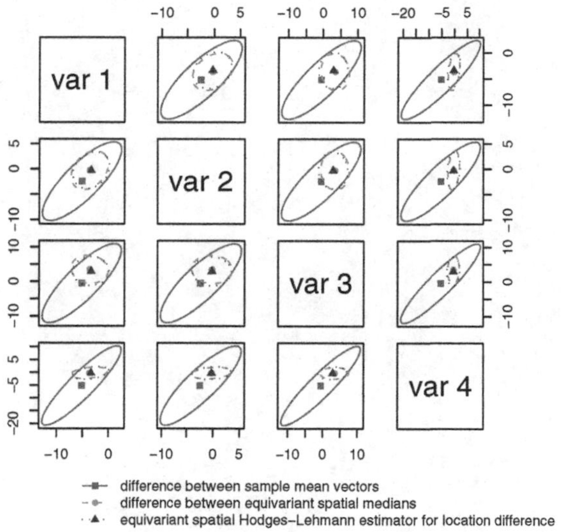

Fig. 11.5 Egyptian skulls data with an outlier. Estimates of the difference between the first and last epochs.

for SAS macros for spatial sign MANOVA methods. Puri and Sen (1971) give a full description of the several-sample location tests based on the vector of marginal ranks. Chakraborty and Chaudhuri (1999) and Nordhausen et al. (2006) propose and consider invariant versions of Puri-Sen tests.

Randles and Peters (1990), Peters and Randles (1990b), and Randles (1992) use interdirections in the test constructions. The tests based on data depth are given in Liu (1992), Liu and Singh (1993), and Liu et al. (1999). Multivariate Oja signs and ranks are used in Hettmansperger and Oja (1994), Hettmansperger et al. (1998) and Oja (1999).

Chapter 12
Randomized blocks

Abstract A multivariate extension of the Friedman test which is based on the spatial ranks is discussed. Related adjusted and unadjusted treatment effect estimates are considered as well. Again, the test using outer standardization is rotation invariant but unfortunately not invariant under heterogeneous scaling of the components. Invariant (equivariant) versions of the test (estimates) based on inner standardization are discussed as well.

12.1 The problem and the test statistics

The blocked design for the comparison of $c \geq 2$ treatments is obtained by generalizing the paired-sample design. A randomized complete block design with n blocks then requires blocks of equal size c; the c subjects in each block are randomly assigned to all c treatments. For univariate response variables, the most popular tests for considering the null hypothesis of no treatment differences are the regular balanced two-way analysis of variance (ANOVA) test and the nonparametric and (under the null hypothesis) distribution-free rank-based Friedman test. These tests are thus based on the identity score and on the rank score, respectively. Naturally, the permutation test version of the ANOVA test is conditionally distribution-free as well. For a discussion of the univariate tests and corresponding treatment difference estimates, see Hollander and Wolfe (1999) or Lehmann (1998).

We discuss a p-variate generalization of the Friedman test with the companion treatment effect estimates. The results and discussion here are based on Möttönen et al. (2003). We do not use the spatial sign score function here as the limiting properties of the test in that case seem too complicated.

The design and the data. The data consist of $N = nc$ p-dimensional vectors. The $N = nc$ subjects are in n blocks of equal size and within each block the c subjects are assigned to c treatments at random. The p-variate observations are then usually given in an $n \times k$ table as follows.

H. Oja, *Multivariate Nonparametric Methods with R: An Approach Based on Spatial Signs and Ranks*, Lecture Notes in Statistics 199, DOI 10.1007/978-1-4419-0468-3_12,

	Treatments			
Blocks	1	2	$\cdots$	c
1	$\mathbf{y}_{11}$	$\mathbf{y}_{12}$	$\cdots$	$\mathbf{y}_{1c}$
2	$\mathbf{y}_{21}$	$\mathbf{y}_{22}$	$\cdots$	$\mathbf{y}_{2c}$
$\vdots$	$\vdots$	$\vdots$	$\ddots$	$\vdots$
n	$\mathbf{y}_{n1}$	$\mathbf{y}_{n2}$	$\cdots$	$\mathbf{y}_{nc}$

The $N \times p$ data matrix can also be written as

$$\mathbf{Y} = \begin{pmatrix} \mathbf{Y}_1 \\ \ldots \\ \mathbf{Y}_n \end{pmatrix},$$

where blockwise $c \times p$ data matrices are

$$\mathbf{Y}_i = (\mathbf{y}_{i1}, \ldots, \mathbf{y}_{ic})', \quad i = 1, \ldots, n.$$

Distributional assumptions. Due to the randomization, it is natural to assume that

$$\mathbf{Y}_i = \mu + \mathbf{1}_c \theta_i' + \varepsilon_i, \quad i = 1, \ldots, n,$$

where θ_i, $i = 1, \ldots, n$, is the block effect (p-vector), and

$$\mu = (\mu_1, \ldots, \mu_c)'$$

is the $c \times p$ matrix of the treatment effects, $\Sigma_{i=1}^{c} \mu_i = \mathbf{0}$, and the rows of the $c \times p$ random matrix ε_i are dependent but exchangeable; that is,

$$\mathbf{P}\varepsilon_i \sim \varepsilon_i, \; i = 1, \ldots, n,$$

for all $c \times c$ permutation matrices $\mathbf{P}$. We also assume that $\varepsilon_1, \ldots, \varepsilon_n$ are independent.

Problem. We wish to test the hypothesis of no treatment effects; that is,

$$H_0 : \mu_1 = \cdots = \mu_c = \mathbf{0}.$$

As in the several-independent-sample case, we also wish to estimate the differences of the treatment effects; that is,

$$\Delta_{jj'} = \mu_{j'} - \mu_j, \quad j, j' = 1, \ldots, c.$$

Note that, under the null hypothesis,

$$\mathbf{PY}_i \sim \mathbf{Y}_i, \quad \text{for all } c \times c \text{ permutation matrices } \mathbf{P}.$$

Under the null hypothesis and under a stronger assumption that $\varepsilon_1,...,\varepsilon_n$ are independent and identically distributed, the random matrices $\mathbf{Y}_i$ are also independent and identically distributed (and the limiting distribution of the test statistic given later can be easily found).

Classical MANOVA test. We first use the identity score in the test construction. The first step is to center the observed values in each block, that is,

$$\mathbf{y}_{ij} \rightarrow \hat{\mathbf{y}}_{ij} = \mathbf{y}_{ij} - \bar{\mathbf{y}}_i, \quad i = 1,...,n; \ j = 1,...,c,$$

where $\bar{\mathbf{y}}_i = (1/c)\sum_{j=1}^{c} \mathbf{y}_{ij}$, Next write

$$\hat{\mathbf{y}}_{.j} = \sum_{i=1}^{n} \hat{\mathbf{y}}_{ij}, \quad j = 1,...,c$$

and

$$\hat{\mathbf{B}} = \frac{1}{nc} \sum_{i=1}^{n} \sum_{j=1}^{c} \hat{\mathbf{y}}_{ij} \hat{\mathbf{y}}_{ij}',$$

which is just the covariance matrix estimate for the within-blocks variation.

Now we can define a squared form test statistic for testing H_0 and give its limiting permutational distribution.

Definition 12.1. MANOVA test statistic for testing H_0 is

$$Q^2 = Q^2(\mathbf{Y}) = \frac{c-1}{nc} \sum_{j=1}^{c} \hat{\mathbf{y}}_{.j}' \hat{\mathbf{B}}^{-1} \hat{\mathbf{y}}_{.j}.$$

Asymptotically distribution-free MANOVA test. Möttönen et al. (2003) showed that if $(\mathbf{Y}_i)$ is a (constant) sequence of matrices with uniformly bounded second and third moments and if $\hat{\mathbf{B}} \rightarrow \mathbf{B} > 0$, then the limiting permutation null distribution of $Q^2(\mathbf{Y})$ is a $\chi^2_{p(c-1)}$ distribution. If $\mathbf{Y}_1,...,\mathbf{Y}_n$ are i.i.d with bounded second moments then the limiting (unconditional) distribution is again the same $\chi^2_{p(k-1)}$. (As the test is based on the centered observations, the same limit is obtained even if the $\mathbf{Y}_i - \mathbf{1}_c\theta_i'$ are i.i.d for some p-vectors θ_i, $i = 1,2,...$. The constants θ_i, $i = 1,...,n$, are then the fixed block effects.)

Note also that if

$$\hat{\mathbf{F}} = n \sum_{j=1}^{c} (\bar{\mathbf{y}}_{.j} - \bar{\mathbf{y}}_{..})(\bar{\mathbf{y}}_{.j} - \bar{\mathbf{y}}_{..})'$$

then we can write

$$Q^2 = \frac{c-1}{c} tr(\hat{\mathbf{F}}\hat{\mathbf{B}}^{-1}),$$

which is again the Pillai trace statistic. The statistic Q^2 is naturally affine invariant.

Conditionally distribution-free MANOVA test. If the null hypothesis is true then

$$\begin{pmatrix} \mathbf{Y}_1 \\ \dots \\ \mathbf{Y}_n \end{pmatrix} \sim \begin{pmatrix} \mathbf{P}_1\mathbf{Y}_1 \\ \dots \\ \mathbf{P}_n\mathbf{Y}_n \end{pmatrix}$$

for all $c \times c$ permutation matrices $\mathbf{P}_1,...,\mathbf{P}_n$. The p-value from the permutation test is then given by

$$E_{\mathbf{P}}\left\{I\left(Q^2(\mathbf{PY}) \geq Q^2(\mathbf{Y})\right)\right\},$$

where $\mathbf{P} = diag(\mathbf{P}_1,...,\mathbf{P}_n)$ is uniformly distributed over all $(c!)^n$ possible values.

The multivariate Friedman test. This test can be derived in exactly the same way as the MANOVA tests. The multivariate blockwise centered response vectors $\hat{\mathbf{y}}_{ij}$ are just replaced by multivariate blockwise centered rank vectors $\mathbf{R}_{ij}$. The vector $\mathbf{R}_{ij}$ is thus the centered rank of the observation $\mathbf{y}_{ij}$ among all the observations in the ith block, that is, among $\mathbf{y}_{i1},...,\mathbf{y}_{ic}$, $i = 1,...,n$. The ranks can be displayed in a table as follows.

	Treatments				
Blocks	1	2	...	c	Σ
1	$\mathbf{R}_{11}$	$\mathbf{R}_{12}$	$\cdots$	$\mathbf{R}_{1c}$	$\mathbf{0}$
2	$\mathbf{R}_{21}$	$\mathbf{R}_{22}$	$\cdots$	$\mathbf{R}_{2c}$	$\mathbf{0}$
$\vdots$	$\vdots$	$\vdots$	$\ddots$	$\vdots$	$\vdots$
n	$\mathbf{R}_{n1}$	$\mathbf{R}_{n2}$	$\cdots$	$\mathbf{R}_{nc}$	$\mathbf{0}$
Σ	$\mathbf{R}_{\cdot 1}$	$\mathbf{R}_{\cdot 2}$	$\cdots$	$\mathbf{R}_{\cdot c}$	$\mathbf{0}$

Now write

$$\hat{\mathbf{B}} = \frac{1}{nc} \sum_{i=1}^{n} \sum_{j=1}^{c} \mathbf{R}_{ij} \mathbf{R}'_{ij}.$$

Definition 12.2. The multivariate Friedman test statistic is

$$Q^2 = \frac{c-1}{nc} \sum_{j=1}^{c} \mathbf{R}'_{\cdot j} \hat{\mathbf{B}}^{-1} \mathbf{R}_{\cdot j}.$$

Note that Q^2 is rotation invariant but not invariant under rescaling of the components.

Theorem 12.1. *Assume that the null hypothesis of no-treatment effect is true and that the sequence $(\mathbf{Y}_i)$ is independent and identically distributed up to a location shift. The limiting distribution of Q^2 is a central chisquare distribution with $p(c-1)$ degrees of freedom.*

Möttönen et al. (2003) also considered the permutation limiting distribution. It is easy to see that the test statistic Q^2 reduces to the classical Friedman test statistic as $p = 1$.

12.2 Limiting distributions and efficiency

For considering asymptotic properties of the tests and estimates we assume that the original observations $\mathbf{y}_{ij}$ are independent and that the cdf of the $\mathbf{y}_{ij}$ is $F(\mathbf{y}-\theta_i-\mu_j)$, where the θ_i are the (possibly random) block effects and the μ_j the treatment effects ($\sum_i \theta_i = \sum_j \mu_j = \mathbf{0}$). We wish to test the hypothesis

$$H_0: \mu_1 = \cdots = \mu_c = \mathbf{0} \quad \text{versus} \quad H_n: \mu_j = \frac{1}{\sqrt{n}}\delta_j, \ j=1,...,c,$$

where also $\sum_j \delta_j = \mathbf{0}$. Let δ denote the vector $(\delta_1' \cdots \delta_{k-1}')'$. As the tests are based on the centered observations and the centered ranks, it is not a restriction to assume in the following that $\theta_1 = \cdots = \theta_n = \mathbf{0}$.

The limiting distribution of the classical MANOVA is then given in the following.

Theorem 12.2. *Under the sequence of alternatives H_n, the limiting distribution of the MANOVA test statistic Q^2 is a noncentral χ^2 distribution with $p(c-1)$ degrees of freedom and a noncentrality parameter*

$$\delta^2_{MANOVA} = \frac{c-1}{c}\sum_{j=1}^{p} \delta_j' \Sigma^{-1} \delta_j,$$

where Σ is the covariance matrix of $\mathbf{y}_{ij}$.

For limiting distributions of the rank tests, we first recall the asymptotic theory for spatial sign and rank tests for comparing two treatments. This is also done to introduce matrices $\mathbf{A}$, $\mathbf{B}_1$, and $\mathbf{B}_2$ needed in the subsequent discussion. First we consider the dependent samples case (matched pairs) and use the one-sample spatial sign test. The one-sample spatial sign test statistic for the difference vectors $\mathbf{y}_{i2}-\mathbf{y}_{i1}$, $i=1,\dots,n$, for example, is defined as

$$\mathbf{T}_{12} = \sum_{i=1}^{n} \mathbf{U}(\mathbf{y}_{i2}-\mathbf{y}_{i1}).$$

As given before in Chapter 8, under H_n,

$$n^{-1/2}\mathbf{T}_{12} \xrightarrow{d} N_d(\mathbf{A}(\delta_2-\delta_1), \mathbf{B}_1),$$

where

$$\mathbf{B}_1 = \mathbf{E}\{\mathbf{U}(\mathbf{y}_{i2}-\mathbf{y}_{i1})\mathbf{U}(\mathbf{y}_{i2}-\mathbf{y}_{i1})'\}$$

is the spatial sign covariance matrix for difference vectors and

$$\mathbf{A} = \mathbf{E}\left\{|\mathbf{y}_{i2}-\mathbf{y}_{i1}|^{-1}\left(\mathbf{I}_p - \mathbf{U}(\mathbf{y}_{i2}-\mathbf{y}_{i1})\mathbf{U}(\mathbf{y}_{i2}-\mathbf{y}_{i1})'\right)\right\}.$$

In the following we also need

$$\mathbf{B}_2 = \mathbf{E}\{\mathbf{U}(\mathbf{y}_{11} - \mathbf{y}_{21})\mathbf{U}(\mathbf{y}_{11} - \mathbf{y}_{31})'\}.$$

Then, under the null hypothesis,

$$\mathbf{B} = \mathbf{E}(\mathbf{R}_{ij}\mathbf{R}_{ij}') = \frac{c-1}{c^2}\mathbf{B}_1 + \frac{(c-1)(c-2)}{c^2}\mathbf{B}_2$$

is the covariance matrix of $\mathbf{R}_{ij}$. (The expected values above are taken under the null hypothesis.)

Now we are ready to give the limiting distributions of the rank tests

Theorem 12.3. *Under the sequence of alternatives H_n, the limiting distribution of the Friedman test statistic Q^2 is a noncentral chi-square distribution with $p(c-1)$ degrees of freedom and a noncentrality parameter*

$$\delta^2_{FRIEDMAN} = \frac{c-1}{c}\sum_{j=1}^{c} \delta_j' \mathbf{A}'\mathbf{B}^{-1}\mathbf{A}\delta_j.$$

Relative efficiencies for comparing the regular MANOVA test and the Friedman test are given by the following theorem.

Theorem 12.4. *The Pitman asymptotic relative efficiency of the multivariate Friedman test with respect to MANOVA is*

$$ARE_{12} = \frac{\delta^2_{FRIEDMAN}}{\delta^2_{MANOVA}} = \frac{\sum_j \delta_j{}'\mathbf{A}'\mathbf{B}^{-1}\mathbf{A}\delta_j}{\sum_j \delta_j{}'\Sigma^{-1}\delta_j}.$$

Table 12.1 lists efficiencies of the Friedman test with respect to the regular MANOVA test. The Friedman test clearly outperforms the classical MANOVA test for heavy-tailed distributions. The efficiencies increase with the dimension p as well as with the number of treatments c. In the multivariate normal case the efficiency goes to one as dimension $p \to \infty$. For $c = 2$, the efficiencies are as the efficiencies of the spatial sign test; the efficiencies go to the efficiencies of the spatial signed-rank test as the number of treatments $c \to \infty$. See Möttönen et al. (2003) for a more detailed study and for the asymptotic relative efficiency of the similarly extended Page test.

12.3 Treatment effect estimates

Consider now pairwise spatial sign test statistics,

$$\mathbf{T}_{jj'} = \sum_{i=1}^{n} \mathbf{U}(\mathbf{y}_{ij'} - \mathbf{y}_{ij}) \quad j, j' = 1, ..., c$$

Table 12.1 Asymptotic relative efficiencies (ARE) of the spatial multivariate Friedman tests with respect to the classical MANOVA in the spherical multivariate $t_{p,\nu}$ distribution case with different choices of p, ν, and the number of treatments c

		c		
p	ν	2	3	10
	4	1.152	1.233	1.367
2	10	0.867	0.925	1.023
	∞	0.785	0.838	0.924
	4	1.245	1.308	1.406
3	10	0.937	0.980	1.048
	∞	0.849	0.887	0.946
	4	1.396	1.427	1.472
10	10	1.050	1.067	1.092
	∞	0.951	0.964	0.981

for comparing treatments j and j'. Note that $\mathbf{T}_{jj}=0$ and $\mathbf{T}_{jj'}=-\mathbf{T}_{j'j}$, $j,j'=1,...,c$. The treatment difference estimates $\hat{\Delta}_{jj'}$ corresponding to $\mathbf{T}_{jj'}$ are the spatial medians of the differences $\mathbf{y}_{ij'}-\mathbf{y}_{ij}$, $i=1,...,n$. Let

$$\hat{\Delta} = (\hat{\Delta}_{jj'})_{j,j'=1,...,c}$$

be the $pc\times c$ matrix of treatment difference estimates ($\hat{\Delta}_{jj}=0,\ j=1,...,k$). As in the univariate and in the case of several independent samples, the problem then is that these estimates may not be consistent in the sense that $\hat{\Delta}_{jj''}=\hat{\Delta}_{jj'}+\hat{\Delta}_{j'j''}$. A solution to this consistency problem is obtained as in the several-independent-sample case. First take the averages

$$\hat{\Delta}_j = -\frac{1}{c}\sum_{j'=1}^{c}\hat{\Delta}_{jj'}.$$

Then the adjusted treatment difference estimates are $\tilde{\Delta}_{jj'}=\hat{\Delta}_{j'}-\hat{\Delta}_j$ and the matrix of these adjusted estimates is then

$$\tilde{\Delta} = (\tilde{\Delta}_{jj'})_{j,j'=1,...,k}.$$

For the univariate case, see Lehmann (1998) and Hollander and Wolfe (1999).

Möttönen et al. (2003) then proved the following.

Theorem 12.5. *Under general assumptions, the limiting distribution of $\sqrt{n}vec(\hat{\Delta}-\Delta)$ is a multivariate singular normal with mean matrix zero and covariance matrix given by*

$$\Sigma_{(ij),(ij)}=\mathbf{A}^{-1}\mathbf{B}_1\mathbf{A}^{-1},\quad \Sigma_{(ij),(il)}=\mathbf{A}^{-1}\mathbf{B}_2\mathbf{A}^{-1},\quad \textit{and}\quad \Sigma_{(ij),(lm)}=\mathbf{0},$$

for $i<j<l<m$, where

$$\Sigma_{(ij),(lm)} = Cov(\hat{\Delta}_{ij}, \hat{\Delta}_{lm}), \quad i,j,l,m = 1,...,c.$$

Moreover, the limiting distribution of $\sqrt{n}vec(\tilde{\Delta} - \Delta)$ *is a multivariate singular normal with mean matrix zero and covariance matrix given by*

$$\Sigma_{(ij),(ij)} = \frac{2c}{c-1}\mathbf{A}^{-1}\mathbf{B}\mathbf{A}^{-1}, \quad \Sigma_{(ij),(il)} = \frac{c}{c-1}\mathbf{A}^{-1}\mathbf{B}\mathbf{A}^{-1}, \quad and \quad \Sigma_{(ij),(lm)} = 0,$$

for $i < j < l < m$*, where now*

$$\Sigma_{(ij),(lm)} = Cov(\tilde{\Delta}_{ij}, \tilde{\Delta}_{lm}), \quad i,j,l,m = 1,...,c.$$

.

The asymptotic relative efficiencies between competing treatment difference estimates is usually given in terms of the Wilks generalized variance, that is, the determinant of the covariance matrix. The asymptotic relative efficiency (ARE) of $\tilde{\Delta}_{jj'}$ with respect to $\hat{\Delta}_{jj'}$ is then given by

$$\left(\frac{\det(\mathbf{A}^{-1}\mathbf{B}_1\mathbf{A}^{-1})}{\det\left(\frac{2c}{c-1}\mathbf{A}^{-1}\mathbf{B}\mathbf{A}^{-1}\right)} \right)^{1/p}.$$

On the other hand, the asymptotic relative efficiency of $\tilde{\Delta}_{jj'}$ with respect to $\bar{\mathbf{y}}_{\cdot j'} - \bar{\mathbf{y}}_{\cdot j}$ is

$$\left(\frac{\det(\Sigma)}{\det(\mathbf{A}^{-1}\mathbf{B}\mathbf{A}^{-1})} \right)^{1/p}.$$

Note that when the observations come from a spherical distribution the asymptotic relative efficiency of the estimate $\tilde{\Delta}_{jj'}$ is the same as the asymptotic relative efficiency of the multivariate Friedman test in Theorem 12.4.

Table 12.2 Asymptotic relative efficiency of $\tilde{\Delta}_{jj'}$ with respect to $\hat{\Delta}_{jj'}$ in the spherical multivariate normal distribution case with different choices of dimension p and with different numbers of treatments c

	number of treatments c		
p	2	3	10
2	1.000	1.067	1.176
3	1.000	1.045	1.114
10	1.000	1.013	1.032

12.4 Affine invariant tests and affine equivariant estimates

The tests and estimates discussed above are rotation invariant/equivariant but unfortunately not scale invariant/equivariant. Due to the lack of the scale invariance property, rescaling of one of the response variables, for example, change the results (p-values) and also highly reduce the efficiency of the tests and estimates. The transformation and retransformation approach introduced by Chakraborty and Chaudhuri (1996) may again be used to construct affine invariant/equivariant versions of the tests and estimates.

To repeat, the idea in the transformation and retransformation approach is as follows. Let $\mathbf{S} = \mathbf{S}(\mathbf{Y})$ be any scatter matrix (estimate of Σ) based on the data. Transform your data set,

$$\hat{\mathbf{Y}} = \mathbf{Y}\mathbf{S}^{-1/2},$$

and construct the test based on the transformed data $\hat{\mathbf{Y}}$. Then find the treatment difference estimates $\hat{\Delta}_{jj'}(\hat{\mathbf{Y}})$ for the transformed data and retransform the estimates:

$$\tilde{\Delta}_{jj'}(\mathbf{Y}) = \mathbf{S}^{1/2}\hat{\Delta}_{jj'}(\hat{\mathbf{Y}}).$$

12.5 Examples and final remarks

We now illustrate the use of the multivariate Friedman test on a dataset earlier analyzed by Seber (1984). A randomized complete block design experiment was arranged to study the effects of six different treatments on plots of bean plants infested by the serpentine leaf miner insect. In this study the number of treatments was $c = 6$ and the number of blocks was $n = 4$. The measurement vectors consist of three different variables: y_1 is the number of miners per leaf, y_2 is the weight of beans per plot (in kilograms) and $y_3 = \sin^{-1}(\sqrt{pr})$, where pr is the proportion of leaves infested with borer. See Table 12.3 and Figure 12.1 for the original dataset. The blockwise centered ranks are given in Table 12.4.

The observed value of the multivariate Friedman test statistic Q_r^2 is now 30.48 and using χ^2_{15} distribution the corresponding p-value is approximately 0.01. The results are similar for the affine invariant version of the test (inner standardization). For the MANOVA test the standardized test statistic and p-value are 32.10 and 0.006, respectively. Also estimated p-values for the exact permutation tests are given in the following printout.

```
> data(beans)
> plot(beans)
```

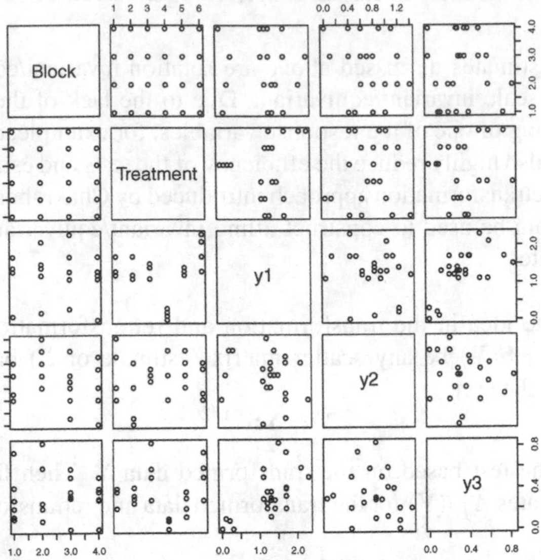

Fig. 12.1 Bean plants data.

Table 12.3 Bean plants data

	Treatments					
Blocks	1	2	3	4	5	6
	1.7	1.7	1.4	0.1	1.3	1.7
1	0.4	1.0	0.8	0.8	1.0	0.5
	0.20	0.40	0.28	0.10	0.12	0.74
	1.2	1.2	1.5	0.2	1.4	2.1
2	1.4	0.6	0.8	1.2	1.2	1.0
	0.20	0.25	0.83	0.08	0.20	0.59
	1.3	1.7	1.1	0.3	1.3	2.3
3	0.6	0.1	0.7	1.2	0.8	0.4
	0.36	0.32	0.58	0.00	0.30	0.50
	1.7	1.1	1.1	0.0	1.2	1.3
4	1.1	0.0	0.9	0.4	0.6	0.9
	0.39	0.29	0.50	0.00	0.36	0.28

```
> Y<-cbind(beans$y1,beans$y2,beans$y3)

> mv.2way.test(Y, beans$Block, beans$Treatment,
               score="r", stand="o", method="a")

        Multivariate Friedman test using spatial ranks
```

```
data:  Y by beans$Treatment within beans$Block
Q2 = 30.4843, df = 15, p-value = 0.01029
alternative hypothesis: true location difference
  between some groups is not equal to c(0,0,0)
> mv.2way.test(Y, beans$Block, beans$Treatment, "r", "i", "a")

        Affine invariant multivariate Friedman test
        using spatial ranks

data:  Y by beans$Treatment within beans$Block
Q2 = 30.7522, df = 15, p-value = 0.00948
alternative hypothesis: true location difference
  between some groups is not equal to c(0,0,0)
> mv.2way.test(Y, beans$Block, beans$Treatment, "i", "o", "a")

        MANOVA test in a randomized complete block design

data:  Y by beans$Treatment within beans$Block
Q2 = 32.1024, df = 15, p-value = 0.006235
alternative hypothesis: true location difference
  between some groups is not equal to c(0,0,0)

> mv.2way.test(Y, beans$Block, beans$Treatment, "r", "o", "p")

        Multivariate Friedman test using spatial ranks

data:  Y by beans$Treatment within beans$Block
Q2 = 30.4843, replications = 1000, p-value < 2.2e-16
alternative hypothesis: true location difference
  between some groups is not equal to c(0,0,0)

> mv.2way.test(Y, beans$Block, beans$Treatment, "i", "o", "p")

        MANOVA test in a randomized complete block design

data:  Y by beans$Treatment within beans$Block
Q2 = 32.1024, replications = 1000, p-value = 0.002
alternative hypothesis: true location difference
  between some groups is not equal to c(0,0,0)
```

12.6 Other approaches

Möttönen et al. (2003) gave an extension of the Page test as well. Another possibility is to use the marginal centered ranks as described in Puri and Sen (1971). This approach is scale invariant/equivariant but not rotation invariant/equivariant, and the efficiency can again be really poor if the response variables are highly correlated. The transformation and retransformation approach may be used to construct affine

Table 12.4 Bean Plants Data: Spatial Ranks

	Treatments						
Blocks	1	2	3	4	5	6	Sum
	0.35	0.43	-0.08	-0.81	-0.20	0.31	0.00
1	-0.50	0.40	0.00	0.02	0.38	-0.31	0.00
	-0.21	0.14	-0.03	-0.13	-0.32	0.56	0.00
	-0.15	-0.13	0.14	-0.77	0.17	0.73	0.00
2	0.49	-0.52	-0.24	0.11	0.17	-0.01	0.00
	-0.17	-0.15	0.53	-0.18	-0.21	0.18	0.00
	-0.02	0.28	-0.35	-0.69	0.01	0.77	0.00
3	-0.13	-0.56	0.08	0.38	0.35	-0.12	0.00
	-0.03	-0.07	0.39	-0.25	-0.17	0.12	0.00
	0.65	0.01	-0.18	-0.75	0.06	0.21	0.00
4	0.45	-0.69	0.32	-0.16	-0.23	0.30	0.00
	0.07	-0.03	0.32	-0.22	0.03	-0.17	0.00
	0.83	0.60	-0.47	-3.01	0.04	2.02	0.00
Sum	0.31	-1.37	0.17	0.36	0.67	-0.14	0.00
	-0.35	-0.11	1.22	-0.78	-0.67	0.69	0.00

invariant/equivariant (and consequently more efficient) versions of the tests and estimates. A third possibility is to use affine equivariant centered ranks based on the Oja criterion; see Oja (1999). To compute blockwise centered ranks one has to require, however, that $c \geq p+1$, which may be a serious limitation in practice.

Chapter 13
Multivariate linear regression

Abstract In this chapter we consider the multivariate multiple regression problem. The tests and estimates are again based on identity, spatial sign, and spatial rank scores. The estimates obtained in this way are then the regular LS estimate, the LAD estimate based on the mean deviation of the residuals (from the origin) and the estimate based on mean difference of the residuals. The estimates are thus multivariate extensions of the univariate L_1 estimates. Equivariant/invariant versions are found using inner centering and standardization.

13.1 General strategy

Assume that $(\mathbf{X}, \mathbf{Y})$ is the data matrix and we consider the linear regression model

$$\mathbf{Y} = \mathbf{X}\beta + \varepsilon,$$

where $\mathbf{Y} = (\mathbf{y}_1, ..., \mathbf{y}_n)'$ is an $n \times p$ matrix of n observed values of p response variables, $\mathbf{X} = (\mathbf{x}_1, ..., \mathbf{x}_n)'$ is an $n \times q$ matrix of observed values of q explaining variables, β is an $q \times p$ matrix of regression coefficients, and $\varepsilon = (\varepsilon_1, ..., \varepsilon_n)'$ is an $n \times p$ matrix of residuals. One can then also write

$$\mathbf{y}_i = \beta' \mathbf{x}_i + \varepsilon_i, \quad i = 1, ..., n.$$

We assume that $\varepsilon = (\varepsilon_1, ..., \varepsilon_n)'$ is a random sample from a p-variate distribution "centered" at the origin. Throughout this chapter we assume the following.

Assumption 6

$$\frac{1}{n}\mathbf{X}'\mathbf{X} \to \mathbf{D} \quad \textit{and} \quad \frac{\max_{1 \le i \le n}\{\mathbf{x}_i' \mathbf{C}' \mathbf{C} \mathbf{x}_i\}}{\sum_{i=1}^n \{\mathbf{x}_i' \mathbf{C}' \mathbf{C} \mathbf{x}_i\}} \to 0$$

H. Oja, *Multivariate Nonparametric Methods with R: An Approach Based on Spatial Signs and Ranks*, Lecture Notes in Statistics 199, DOI 10.1007/978-1-4419-0468-3_13,

for some positive definite $q \times q$ matrix $\mathbf{D}$ and for all $p \times q$ matrices $\mathbf{C}$ with positive rank.

Testing problem I. Consider first the problem of testing the null hypothesis $H_0:\ \beta = \mathbf{0}$. The null hypothesis thus simply says that $\mathbf{y}_1, ..., \mathbf{y}_n$ are independent and identically distributed with a joint distribution centered at the origin. By a centered observation, we mean here that $\mathbf{E}\{\mathbf{T}(\mathbf{y}_i)\} = \mathbf{0}$ for the chosen p-variate score function $\mathbf{T}(\mathbf{y})$. We write

$$\mathbf{T}_i = \mathbf{T}(\mathbf{y}_i) \quad \text{and} \quad \mathbf{T}_i(\beta) = \mathbf{T}(\mathbf{y}_i - \beta'\mathbf{x}_i), \quad i = 1, ..., n,$$

and

$$\mathbf{T} = (\mathbf{T}_1, ..., \mathbf{T}_n)' \quad \text{and} \quad \mathbf{T}(\beta) = (\mathbf{T}_1(\beta), ..., \mathbf{T}_n(\beta))'.$$

Again write $\mathbf{A} = \mathbf{E}\{\mathbf{T}(\varepsilon_i)\mathbf{L}(\varepsilon_i)'\}$ and $\mathbf{B} = \mathbf{E}\{\mathbf{T}(\varepsilon_i)\mathbf{T}(\varepsilon_i)'\}$. As before, $\mathbf{L}(\mathbf{y})$ is the optimal multivariate location score function. If $\mathbf{B}$ exists, then under the null hypothesis and under our design Assumption 6,

$$n^{-1/2}\text{vec}(\mathbf{T}'\mathbf{X}) \;=\; n^{-1/2}(\mathbf{X}' \otimes \mathbf{I}_p)\text{vec}(\mathbf{T}')$$

has a limiting $N_{pq}(\mathbf{0}, \mathbf{D} \otimes \mathbf{B})$ distribution and then the test statistic using the outer standardization

$$Q^2 = Q^2(\mathbf{X}, \mathbf{Y}) = n \cdot tr((\mathbf{T}'\mathbf{P}_\mathbf{X}\mathbf{T})(\mathbf{T}'\mathbf{T})^{-1}) \;\to_d\; \chi^2_{pq},$$

where, as before, $\mathbf{P}_\mathbf{X} = \mathbf{X}(\mathbf{X}'\mathbf{X})^{-1}\mathbf{X}'$ is an $n \times n$ projection matrix to the linear space spanned by the columns of $\mathbf{X}$. The test thus compares the covariance matrix of the projected and transformed data to the covariance matrix of the transformed data. Under the sequence of alternatives $H_n:\ \beta = n^{-1/2}\delta$ where a $q \times p$ matrix δ gives the direction of the alternative sequence, the limiting distribution of $n^{-1/2}\text{vec}(\mathbf{T}'\mathbf{X})$ is often a $N_{pq}(\text{vec}(\mathbf{A}\delta'\mathbf{D}), \mathbf{D} \otimes \mathbf{B})$ distribution, and that of Q^2 is a noncentral chi-square distribution with pq degrees of freedom and noncentrality parameter

$$\text{vec}(\delta')'(\mathbf{D} \otimes (\mathbf{A}\mathbf{B}^{-1}\mathbf{A}))\text{vec}(\delta') = tr\left((\delta'\mathbf{D}\delta)(\mathbf{A}\mathbf{B}^{-1}\mathbf{A})\right).$$

The distribution of the test statistic at the true value β (close to the origin) can then be approximated by a noncentral chi-square distribution with pq degrees of freedom and noncentrality parameter

$$tr\left((\Delta'\mathbf{P}_\mathbf{X}\Delta)\left(\mathbf{A}\mathbf{B}^{-1}\mathbf{A}\right)\right),$$

where $\Delta = \mathbf{X}\beta$.

If one uses the inner standardization, one first finds a full rank transformation matrix $\mathbf{S}^{-1/2}$ such that, if we transform

$$\mathbf{y}_i \to \hat{\mathbf{T}}_i = \mathbf{T}(\mathbf{S}^{-1/2}\mathbf{y}_i) \quad \text{and} \quad \mathbf{Y} \to \hat{\mathbf{T}} = (\hat{\mathbf{T}}_1, ..., \hat{\mathbf{T}}_n)'$$

then

$$p \cdot \hat{\mathbf{T}}'\hat{\mathbf{T}} = tr(\hat{\mathbf{T}}'\hat{\mathbf{T}})\mathbf{I}_p.$$

The test statistic is then

$$Q^2 = Q^2(\mathbf{X},\mathbf{Y}) = n \cdot tr((\hat{\mathbf{T}}'\mathbf{P_X}\hat{\mathbf{T}})(\hat{\mathbf{T}}'\hat{\mathbf{T}})^{-1}) = \frac{np}{tr(\hat{\mathbf{T}}'\hat{\mathbf{T}})}|\mathbf{P_X}\hat{\mathbf{T}}|^2$$

which also often has the limiting χ^2_{pq} null distribution. The p value from a conditionally distribution-free permutation test is obtained as

$$\mathbf{E_P}\left\{I\left(Q^2(\mathbf{PX},\mathbf{Y}) \geq Q^2(\mathbf{X},\mathbf{Y})\right)\right\},$$

where $\mathbf{P}$ is uniformly distributed in the space of all possible $n \times n$ permutation matrices.

Estimation problem. We consider the linear regression model

$$\mathbf{Y} = \mathbf{X}\beta + \varepsilon$$

and wish to estimate unknown β. The estimate that is based on the score function $\mathbf{T}$ then often solves

$$\mathbf{T}(\hat{\beta})'\mathbf{X} = \mathbf{0}.$$

We thus find the estimate $\hat{\beta}$ such that the transformed estimated residuals are uncorrelated with the explaining variable. Note that usually one of the explaining variables in $\mathbf{X}$ corresponds to the intercept term in the regression model (the corresponding column in $\mathbf{X}$ is $\mathbf{1}_n$); this implies that the transformed residuals also sum up to zero.

Under general assumptions, the approximate connection between the estimate and test is

$$\sqrt{n}(\hat{\beta}-\beta)' = \mathbf{A}^{-1}\left(\frac{1}{\sqrt{n}}\mathbf{T}(\beta)'\mathbf{X}\right)\mathbf{D}^{-1} + \mathbf{o}_P(1)$$

which further implies that

$$\sqrt{n}\,\mathrm{vec}\left((\hat{\beta}-\beta)'\right) \to_d N_{qp}(\mathbf{0}, \mathbf{D}^{-1} \otimes (\mathbf{A}^{-1}\mathbf{B}\mathbf{A}^{-1})),$$

where, as before, $\mathbf{A} = \mathbf{E}\{\mathbf{T}(\varepsilon_i)\mathbf{L}(\varepsilon_i)'\}$ and $\mathbf{B} = \mathbf{E}\{\mathbf{T}(\varepsilon_i)\mathbf{T}(\varepsilon_i)'\}$. Recall that $\mathbf{L}$ is the optimal score function. If one uses inner standardization in the estimation problem, one first finds an estimate $\hat{\beta}$ and a full rank transformation matrix $\mathbf{S}^{-1/2}$ such that, if we transform

$$\mathbf{y}_i \to \hat{\mathbf{T}}_i = \mathbf{T}(\mathbf{S}^{-1/2}(\mathbf{y}_i - \hat{\beta}'\mathbf{x}_i)) \quad \text{and} \quad \mathbf{Y} \to \hat{\mathbf{T}} = (\hat{\mathbf{T}}_1, ..., \hat{\mathbf{T}}_n)'$$

then

$$\hat{\mathbf{T}}'\mathbf{X} = \mathbf{0} \quad \text{and} \quad p \cdot \hat{\mathbf{T}}'\hat{\mathbf{T}} = tr(\hat{\mathbf{T}}'\hat{\mathbf{T}})\mathbf{I}_p.$$

Testing problem II. Consider next the partitioned linear regression model

$$\mathbf{Y} = \mathbf{X}_1\beta_1 + \mathbf{X}_2\beta_2 + \varepsilon$$

where $\mathbf{X}_1$ (resp., $\mathbf{X}_2$) is an $n \times q_1$ (resp., $n \times q_2$) matrix. We wish to test the null hypothesis $H_0: \ \beta_2 = \mathbf{0}$.

To construct the test statistic, first find the centered scores (the centering under the null hypothesis)

$$\hat{\mathbf{T}} = \mathbf{T}(\hat{\beta}_1, \mathbf{0})$$

such that $\hat{\mathbf{T}}'\mathbf{X}_1 = \mathbf{0}$. We also write $\hat{\mathbf{X}}_2 = (\mathbf{I}_n - \mathbf{P}_{\mathbf{X}_1})\mathbf{X}_2$. Then the test statistic

$$Q^2 = Q^2(\mathbf{X}_1, \mathbf{X}_2, \mathbf{Y}) = n \cdot tr\left(\hat{\mathbf{T}}'\mathbf{P}_{\hat{\mathbf{X}}_2}\hat{\mathbf{T}}(\hat{\mathbf{T}}'\hat{\mathbf{T}})^{-1}\right)$$

has an approximate chi-square distribution with $q_2 p$ degrees of freedom. The distribution of the test statistic for $\Delta = \mathbf{X}\beta$ (close to the null point where $\mathbf{P}_{\mathbf{X}_1}\Delta = \Delta$) can now be approximated by a noncentral chi-square distribution with $q_2 p$ degrees of freedom and noncentrality parameter

$$tr\left((\hat{\Delta}'\mathbf{P}_{\mathbf{X}_2}\hat{\Delta})\,(\mathbf{A}\mathbf{B}^{-1}\mathbf{A}) \right).$$

where $\hat{\Delta} = (\mathbf{I}_n - \mathbf{P}_{\mathbf{X}_1})\Delta$.

Note also that, under general assumptions,

$$\sqrt{n}(\hat{\beta}_2 - \beta_2)' = \mathbf{A}^{-1}\left(\frac{1}{\sqrt{n}}\mathbf{T}(\beta)'\hat{\mathbf{X}}_2\right)\left(\frac{1}{n}\hat{\mathbf{X}}_2'\hat{\mathbf{X}}_2\right)^{-1} + o_P(1)$$

and then the *Wald test statistic*

$$n\ \mathrm{vec}(\hat{\beta}_2)'\widehat{\mathbf{COV}(\hat{\beta}_2)}^{-1}\ \mathrm{vec}(\hat{\beta}_2),$$

where

$$\widehat{\mathbf{COV}(\hat{\beta}_2)} = \left(\frac{1}{n}\hat{\mathbf{X}}_2{}'\hat{\mathbf{X}}_2\right)^{-1} \otimes (\hat{\mathbf{A}}^{-1}\hat{\mathbf{B}}\hat{\mathbf{A}}^{-1})$$

(with consistent estimates of $\hat{\mathbf{A}}$ and $\hat{\mathbf{B}}$) is asymptotically equivalent with Q^2.

If one also uses inner standardization (to attain affine invariance), one first transforms

$$\mathbf{y}_i \rightarrow \hat{\mathbf{T}}_i = \mathbf{T}(\mathbf{S}^{-1/2}(\mathbf{y}_i - \hat{\beta}_1'\mathbf{x}_{1i})) \quad \text{and} \quad \mathbf{Y} \rightarrow \hat{\mathbf{T}} = (\hat{\mathbf{T}}_1, ..., \hat{\mathbf{T}}_n)'$$

such that

$$\hat{\mathbf{T}}'\mathbf{X}_1 = \mathbf{0} \quad \text{and} \quad p \cdot \hat{\mathbf{T}}'\hat{\mathbf{T}} = tr(\hat{\mathbf{T}}'\hat{\mathbf{T}})\mathbf{I}_p.$$

The test statistic is then the same (but with changed, standardized scores)

$$Q^2 = n \cdot tr\left(\hat{\mathbf{T}}'\mathbf{P}_{\hat{\mathbf{X}}_2}\hat{\mathbf{T}}(\hat{\mathbf{T}}'\hat{\mathbf{T}})^{-1}\right) = \frac{np}{tr(\hat{\mathbf{T}}'\hat{\mathbf{T}})}|\mathbf{P}_{\hat{\mathbf{X}}_2}\hat{\mathbf{T}}|^2.$$

Note that the use of the permutation test is questionable here. It is allowed only if $\mathbf{X}_1$ and $\mathbf{X}_2$ are independent ($\mathbf{X}_2$ gives the treatment in a randomized trial, for example) and then the p value is

$$\mathbf{E}_\mathbf{P}\left\{I\left(Q^2(\mathbf{X}_1,\mathbf{P}\mathbf{X}_2,\mathbf{Y}) \geq Q^2(\mathbf{X}_1,\mathbf{X}_2,\mathbf{Y})\right)\right\}.$$

Again, $\mathbf{P}$ is uniformly distributed in the space of all possible $n \times n$ permutation matrices.

13.2 Multivariate linear L_2 regression

We consider the linear regression model

$$\mathbf{Y} = \mathbf{X}\beta + \varepsilon,$$

where $\mathbf{Y}$ is an $n \times p$ matrix of the response variable, ε is an $n \times p$ matrix of the error variable, $\mathbf{X}$ is an $n \times q$ matrix of explaining variables, and β is a $q \times p$ matrix of regression coefficients. For each individual i, we then have

$$\mathbf{y}_i = \beta'\mathbf{x}_i + \varepsilon_i, \quad i = 1,...,n.$$

In this section we use the identical score function $\mathbf{T}(\mathbf{y}) = \mathbf{y}$ and therefore assume that ε is a random sample from a p-variate distribution with

$$\mathbf{E}(\varepsilon_i) = \mathbf{0} \quad \text{and} \quad \mathbf{COV}(\varepsilon_i) = \Sigma.$$

We thus need the assumption that the second moments exist. We assume that $\mathbf{X}'\mathbf{X}$ is a full-rank matrix, the rank is q, and thus it has the inverse. For the asymptotical results we assume that the explaining (design) variables are fixed and satisfy Assumption 6 with

$$\frac{1}{n}\mathbf{X}'\mathbf{X} \to \mathbf{D} \quad \text{as } n \to \infty,$$

where the rank of $\mathbf{D}$ is also q. We naturally often assume that the first column of $\mathbf{X}$ is $\mathbf{1}_n$ so that the first row of β is the so-called *intercept parameter*. Note that the one-sample and several-sample location problems are special cases here.

Testing problem I. We first consider the problem of testing the null hypothesis $H_0: \ \beta = \mathbf{0}$ versus $H_1: \ \beta \neq \mathbf{0}$. We thus wish to test whether there is any linear structure in the population. The null hypothesis simply says that $\mathbf{E}(\mathbf{y}_i) = \mathbf{0}$ for all i and therefore is independent on the values $\mathbf{x}_i$, $i = 1,...,n$. If we use the identity score, then the test statistic is simply

$$\mathbf{Y}'\mathbf{X} \quad \text{or equivalently} \quad \mathbf{Y}'\mathbf{X}(\mathbf{X}'\mathbf{X})^{-1}.$$

Under our assumptions and under the null hypothesis,

$$\frac{1}{n}\mathbf{X}'\mathbf{X} \to \mathbf{D} \quad \text{and} \quad \frac{1}{n}\mathbf{Y}'\mathbf{Y} \to_P \Sigma$$

and therefore

$$n^{-1/2}\text{vec}(\mathbf{Y}'\mathbf{X}) \to_d N_{pq}(\mathbf{0}, \mathbf{D} \otimes \Sigma)$$

and finally

$$Q^2 = Q^2(\mathbf{Y}) = n \cdot tr(\mathbf{Y}'\mathbf{P}_\mathbf{X}\mathbf{Y}(\mathbf{Y}'\mathbf{Y})^{-1}) \to_d \chi^2_{pq}.$$

This test is affine invariant in the sense that if we transform

$$(\mathbf{X}, \mathbf{Y}) \quad \to \quad (\mathbf{X}\mathbf{V}, \mathbf{Y}\mathbf{W}),$$

where $\mathbf{V}$ and $\mathbf{W}$ are $q \times q$ and $p \times p$ full rank transformation matrices, then the value of the test statistic remains unchanged; that is,

$$Q^2(\mathbf{X}, \mathbf{Y}) = Q^2(\mathbf{X}\mathbf{V}, \mathbf{Y}\mathbf{W}).$$

Estimation problem. In the L_2 estimation , the estimate $\hat{\beta}$ minimizes

$$D_n(\beta) = \frac{1}{n}|\mathbf{Y} - \mathbf{X}\beta|^2 = \frac{1}{n}tr\left((\mathbf{Y} - \mathbf{X}\beta)'(\mathbf{Y} - \mathbf{X}\beta)\right) = \mathbf{AVE}\{|\mathbf{y}_i - \beta'\mathbf{x}_i|^2\}$$

or solves

$$(\mathbf{Y} - \mathbf{X}\hat{\beta})'\mathbf{X} = \mathbf{0}.$$

If $\mathbf{X}$ has rank q, then the solution is

$$\hat{\beta} = (\mathbf{X}'\mathbf{X})^{-1}\mathbf{X}'\mathbf{Y},$$

and the estimate is also called the *least squares (LS)* estimate. As

$$\hat{\beta} - \beta = (\mathbf{X}'\mathbf{X})^{-1}\mathbf{X}'(\mathbf{Y} - \mathbf{X}\beta) = (\mathbf{X}'\mathbf{X})^{-1}\mathbf{X}'\varepsilon$$

we easily get that

$$\sqrt{n}\ \text{vec}((\hat{\beta} - \beta)') \to_d N_{qp}(\mathbf{0}, \mathbf{D}^{-1} \otimes \Sigma).$$

The estimate $\hat{\beta} = \hat{\beta}(\mathbf{X}, \mathbf{Y})$ has the following important equivariance properties.

1. *Regression equivariance:*

$$\hat{\beta}(\mathbf{X}, \mathbf{X}\mathbf{H} + \mathbf{Y}) = \hat{\beta}(\mathbf{X}, \mathbf{Y}) + \mathbf{H} \quad \text{for all full-rank matrices } \mathbf{H}.$$

2. $\mathbf{Y}$ *equivariance:*

$$\hat{\beta}(\mathbf{X}, \mathbf{YW}) = \hat{\beta}(\mathbf{X}, \mathbf{Y})\mathbf{W} \quad \text{for all full-rank matrices } \mathbf{W}.$$

3. $\mathbf{X}$ *equivariance:*

$$\hat{\beta}(\mathbf{XV}, \mathbf{Y}) = \mathbf{V}^{-1}\hat{\beta}(\mathbf{X}, \mathbf{Y}) \quad \text{for all full-rank matrices } \mathbf{V}.$$

Testing problem II. Next consider the model

$$\mathbf{Y} = \mathbf{X}_1\beta_1 + \mathbf{X}_2\beta_2 + \varepsilon,$$

where the linear part is partitioned into two parts. We wish to test the null hypothesis $H_0 : \beta_2 = 0$. We first center the matrices $\mathbf{Y}$ and $\mathbf{X}_2$ using $\mathbf{X}_1$ (inner centering); that is

$$\mathbf{Y} \to \hat{\mathbf{Y}} = (\mathbf{I}_n - \mathbf{P}_{\mathbf{X}_1})\mathbf{Y} \quad \text{and} \quad \mathbf{X}_2 \to \hat{\mathbf{X}}_2 = (\mathbf{I}_n - \mathbf{P}_{\mathbf{X}_1})\mathbf{X}_2.$$

Then $\hat{\mathbf{Y}}'\mathbf{X}_1 = \mathbf{0}$ and $\hat{\mathbf{X}}_2{}'\mathbf{X}_1 = \mathbf{0}$. The test statistic for testing $H_0 : \beta_2 = 0$ is

$$Q^2 = n \cdot tr\left(\hat{\mathbf{Y}}'\mathbf{P}_{\hat{\mathbf{X}}_2}\hat{\mathbf{Y}}(\hat{\mathbf{Y}}'\hat{\mathbf{Y}})^{-1}\right)$$

which has an approximate chi-square distribution with $q_2 p$ degrees of freedom. This test for $H_0 : \beta_2 = \mathbf{0}$ is affine invariant in the sense that if we transform

$$(\mathbf{X}_1, \mathbf{X}_2, \mathbf{Y}) \quad \to \quad (\mathbf{X}_1\mathbf{V}_1, \mathbf{X}_2\mathbf{V}_2, \mathbf{YW}),$$

where $\mathbf{V}_1$, $\mathbf{V}_2$, and $\mathbf{W}$ are full rank transformation matrices (with ranks q_1, q_2, and p, resp.) then the value of the test statistic is not changed; that is,

$$Q^2(\mathbf{X}_1, \mathbf{X}_2, \mathbf{Y}) = Q^2(\mathbf{X}_1\mathbf{V}_1, \mathbf{X}_2\mathbf{V}_2, \mathbf{YW}).$$

13.3 L_1 regression based on spatial signs

We now assume that, in considering the linear regression model

$$\mathbf{Y} = \mathbf{X}\beta + \varepsilon,$$

ε is a random sample of size n from a p-variate distribution with the spatial median at zero; that is,

$$\mathbf{E}(\mathbf{U}(\varepsilon_i)) = \mathbf{0}.$$

With the spatial sign score we again need the assumption that the density of ε_i is bounded. Again, $\mathbf{X}'\mathbf{X}$ is a full-rank matrix, rank is q, and $\mathbf{X}$ satisfies Assumption 6 with

$$\frac{1}{n}\mathbf{X}'\mathbf{X} \to \mathbf{D} \quad \text{as } n \to \infty.$$

Testing problem I. Consider the testing problem with the null hypothesis H_0 : $\beta = \mathbf{0}$. Under the null hypothesis $\mathbf{E}(\mathbf{U}(\mathbf{y}_i)) = \mathbf{0}$ for all $i = 1,...,n$. If one uses the spatial sign score $\mathbf{U}(\mathbf{y})$, then one first transforms

$$\mathbf{y}_i \to \mathbf{U}_i = \mathbf{U}(\mathbf{y}_i), \quad i = 1,...,n, \quad \text{and} \quad \mathbf{Y} \to \mathbf{U} = (\mathbf{U}_1,...,\mathbf{U}_n)'$$

and the test statistic is based on the covariances between components of $\mathbf{U}$ and $\mathbf{X}$, that is, on the matrix $\mathbf{U}'\mathbf{X}$. Then, under the null hypothesis,

$$n^{-1/2}\text{vec}(\mathbf{U}'\mathbf{X}) \to_d N_{pq}(\mathbf{0}, \mathbf{D} \otimes \mathbf{B}),$$

where $\mathbf{B} = \mathbf{E}\{\mathbf{U}(\varepsilon_i)\mathbf{U}(\varepsilon_i)'\}$ and finally

$$Q^2 = Q^2(\mathbf{Y}) = n \cdot tr(\mathbf{U}'\mathbf{P_X}\mathbf{U}(\mathbf{U}'\mathbf{U})^{-1}) \to_d \chi^2_{pq}.$$

Unfortunately, this test is not affine invariant but an affine invariant test version can be found using inner standardization. Find a transformation matrix $\mathbf{S}^{-1/2}$ such that, if we transform

$$\mathbf{y}_i \to \hat{\mathbf{U}}_i = \mathbf{U}(\mathbf{S}^{-1/2}\mathbf{y}_i) \quad \text{and} \quad \mathbf{Y} \to \hat{\mathbf{U}} = (\hat{\mathbf{U}}_1,...,\hat{\mathbf{U}}_n)'$$

then

$$p \cdot \hat{\mathbf{U}}'\hat{\mathbf{U}} = n\mathbf{I}_p.$$

The transformation is then again Tyler's transformation, and the test statistic is

$$Q^2 = Q^2(\mathbf{X}, \mathbf{Y}) = p \cdot tr(\hat{\mathbf{U}}'\mathbf{P_X}\hat{\mathbf{U}}) = p \cdot |\mathbf{P_X}\hat{\mathbf{U}}|^2.$$

The statistic Q^2 has the limiting χ^2_{pq} null distribution as well.

Estimation problem. In the L_1 estimation based on the spatial sign score, the estimate $\hat{\beta}$ minimizes

$$\mathbf{AVE}\{|\mathbf{y}_i - \beta'\mathbf{x}_i|\} \quad \text{or} \quad D_n(\beta) = \mathbf{AVE}\{|\mathbf{y}_i - \beta'\mathbf{x}_i| - |\mathbf{y}_i|\}.$$

The estimate $\hat{\beta}$ then often solves

$$\mathbf{U}(\hat{\beta})'\mathbf{X} = \mathbf{0},$$

where

$$\mathbf{U}_i(\beta) = \mathbf{U}(\mathbf{y}_i - \beta'\mathbf{x}_i), \quad i = 1,...,n, \quad \text{and} \quad \mathbf{U}(\beta) = (\mathbf{U}_1(\beta),...,\mathbf{U}_n(\beta))'.$$

The estimate is sometimes also called the *least absolute deviation (LAD) estimate*.

The solution $\hat{\beta}$ cannot be given in a closed form but may be easily calculated using the algorithm with the following two iteration steps.

1.
$$\mathbf{e}_i \leftarrow y_i - \beta'\mathbf{x}_i, \quad i = 1, \ldots, n.$$

2.
$$\beta \leftarrow \beta + \left[\mathbf{AVE}\{|\mathbf{e}_i|^{-1}\mathbf{x}_i\mathbf{x}_i'\}\right]^{-1} \mathbf{AVE}\{\mathbf{x}_i\mathbf{U}(\mathbf{e}_i)'\}.$$

Assume for a moment that $\beta = \mathbf{0}$. Then under our assumptions

$$D_n(\beta) \to_P D(\beta) = tr(\mathbf{D}\beta'\mathbf{A}\beta) + o(|\beta|),$$

where $D_n(\beta)$ and $D(\beta)$ are convex. Also

$$nD_n(n^{-1/2}\beta) - \text{vec}(n^{-1/2}\mathbf{U}'\mathbf{X} - \mathbf{A}\beta'\mathbf{D})'\text{vec}(\beta) = o_P(1).$$

See Appendix B. As in the case of the spatial median, it then follows that

$$\sqrt{n}\, \text{vec}((\hat{\beta} - \beta)') \to_d N_{qp}\left(\mathbf{0}, \mathbf{D}^{-1} \otimes (\mathbf{A}^{-1}\mathbf{B}\mathbf{A}^{-1})\right),$$

where

$$\mathbf{A} = \mathbf{E}\{\mathbf{A}(\varepsilon_i)\} \quad \text{and} \quad \mathbf{B} = \mathbf{E}\{\mathbf{B}(\varepsilon_i)\}$$

with

$$\mathbf{A}(\mathbf{y}) = \frac{1}{|\mathbf{y}|}\left[\mathbf{I}_p - \mathbf{U}(\mathbf{y})\mathbf{U}(\mathbf{y})'\right] \text{ and } \mathbf{B}(\mathbf{y}) = \mathbf{U}(\mathbf{y})\mathbf{U}(\mathbf{y})'.$$

Natural consistent estimates of $\mathbf{A}$ and $\mathbf{B}$ are then

$$\hat{\mathbf{A}} = \mathbf{AVE}\left\{\mathbf{A}\left(\mathbf{y}_i - \hat{\beta}'\mathbf{x}_i\right)\right\} \text{ and } \hat{\mathbf{B}} = \mathbf{AVE}\left\{\mathbf{B}\left(\mathbf{y}_i - \hat{\beta}'\mathbf{x}_i\right)\right\},$$

respectively.

The estimate $\hat{\beta} = \hat{\beta}(\mathbf{X}, \mathbf{Y})$ is regression equivariant and $\mathbf{X}$ equivariant but, unfortunately not $\mathbf{Y}$ equivariant. A fully equivariant LAD estimate is again obtained using the transformation retransformation technique. A natural extension of the Hettmansperger-Randles estimate is then obtained with an algorithm that first updates the residuals, then the β matrix, and finally the residual scatter matrix $\mathbf{S}$ as follows.

1.
$$\mathbf{e}_i \leftarrow \mathbf{S}^{-1/2}(\mathbf{y}_i - \beta'\mathbf{x}_i), \quad i = 1, \ldots, n.$$

2.
$$\beta \leftarrow \beta + \left[\mathbf{AVE}\{|\mathbf{e}_i|^{-1}\mathbf{x}_i\mathbf{x}_i'\}\right]^{-1} \mathbf{AVE}\{\mathbf{x}_i\mathbf{U}(\mathbf{e}_i)'\}\mathbf{S}^{1/2}.$$

3.
$$\mathbf{S} \leftarrow p\, \mathbf{S}^{1/2}\, \mathbf{AVE}\{\mathbf{U}(\mathbf{e}_i)\mathbf{U}(\mathbf{e}_i)'\}\, \mathbf{S}^{1/2}.$$

As in the case of the one-sample HR estimate, there is no proof for the convergence of the algorithm but in practice it seems to work. If $\beta = \mathbf{0}$, $(1/n)\mathbf{X}'\mathbf{X} \to \mathbf{D} = \mathbf{I}_p$, and ε_i is spherically distributed around the origin, and the initial regression and shape estimates, say $\mathbf{B}$ and $\mathbf{S}$ are root-n consistent, that is,

$$\sqrt{n}\mathbf{B} = O_P(1) \quad \text{and} \quad \sqrt{n}(\mathbf{S} - \mathbf{I}_p) = O_P(1)$$

with $tr(\mathbf{S}) = p$ then one can again show that the k-step estimates (obtained after k iterations of the above algorithm) satisfy

$$\begin{aligned}\sqrt{n}\mathbf{B}_k = & \left(\frac{1}{p}\right)^k \sqrt{n}\mathbf{B} \\ & + \left[1 - \left(\frac{1}{p}\right)^k\right] \frac{1}{E(r_i^{-1})} \frac{p}{p-1} \sqrt{n}\mathbf{AVE}\{\mathbf{x}_i\mathbf{u}_i'\} + o_P(1)\end{aligned}$$

and

$$\begin{aligned}\sqrt{n}(\mathbf{S}_k - \mathbf{I}_p) = & \left(\frac{2}{p+2}\right)^k \sqrt{n}(\mathbf{S} - \mathbf{I}_p) \\ & + \left[1 - \left(\frac{2}{p+2}\right)^k\right] \frac{p+2}{p} \sqrt{n}\left(p \cdot \mathbf{AVE}\{\mathbf{u}_i\mathbf{u}_i'\} - \mathbf{I}_p\right) + o_P(1).\end{aligned}$$

Asymptotically, the k-step estimate behaves as a linear combination of the initial pair of estimates and the L_1 estimate. The larger k, the more similar is the limiting distribution to that of the L_1 estimate.

Testing problem II. Consider again the model with two parts of explaining variables, $\mathbf{X}_1$ and $\mathbf{X}_2$:

$$\mathbf{Y} = \mathbf{X}_1\beta_1 + \mathbf{X}_2\beta_2 + \varepsilon.$$

We wish to test the null hypothesis that the variables in the $\mathbf{X}_2$ part have no effect on the response variable. Thus $H_0: \ \beta_2 = 0$. In the null case, the estimate of β_1 solves

$$\mathbf{U}(\hat{\beta}_1, \mathbf{0})'\mathbf{X}_1 = \mathbf{0}.$$

Then write

$$\hat{\mathbf{U}} = \mathbf{U}(\hat{\beta}_1, \mathbf{0}) \quad \text{and} \quad \hat{\mathbf{X}}_2 = (\mathbf{I}_n - \mathbf{P}_{\mathbf{X}_1})\mathbf{X}_2.$$

Then $\hat{\mathbf{U}}'\mathbf{X}_1 = \mathbf{0}$ and $\hat{\mathbf{X}}_2'\mathbf{X}_1 = \mathbf{0}$. The test statistic for testing $H_0: \ \beta_2 = \mathbf{0}$ is then

$$Q^2 = n \cdot tr\left(\hat{\mathbf{U}}'\mathbf{P}_{\hat{\mathbf{X}}_2}\hat{\mathbf{U}}(\hat{\mathbf{U}}'\hat{\mathbf{U}})^{-1}\right)$$

with a limiting chi-square distribution with $q_2 p$ degrees of freedom.

Unfortunately, the test statistic Q^2 is not invariant under affine transformations to $\mathbf{Y}$. The affine invariant version of the test is obtained if the spatial sign scores are obtained by using both inner centering and inner standardization. Then the spatial signs

$$\hat{\mathbf{U}}_i = \mathbf{U}\left(\mathbf{S}^{-1/2}(\mathbf{y}_i - \hat{\beta}_1 \mathbf{x}_{1i})\right), \quad i = 1,...,n,$$

satisfy

$$\hat{\mathbf{U}}'\mathbf{X}_1 = \mathbf{0} \quad \text{and} \quad p \cdot \hat{\mathbf{U}}'\hat{\mathbf{U}} = n\mathbf{I}_p.$$

The test statistic is then

$$Q^2 = p \cdot |\mathbf{P}_{\hat{\mathbf{X}}_2} \hat{\mathbf{U}}|^2.$$

13.4 L_1 regression based on spatial ranks

This approach uses the spatial rank function $\mathbf{R}(\mathbf{y})$ as a score function. Note that the spatial ranks are invariant under location shifts. Therefore a separate procedure is needed for the estimation of the intercept vector. This approach here extends the *Wilcoxon-Mann-Whitney* and *Kruskal-Wallis tests* to the general multivariate regression case.

We again assume that

$$\mathbf{Y} = \mathbf{1}_n \mu' + \mathbf{X}\beta + \varepsilon,$$

where $\mathbf{X}$ is an $n \times q$ matrix of genuine explaining variables with regression coefficient matrix β, μ is the intercept, and ε is a random sample from a p-variate continuous distribution. We again need the assumption that the density of ε_i is bounded. Then the densities of $\varepsilon_i - \varepsilon_j$ and $\varepsilon_i + \varepsilon_j$ are also bounded. The design matrix $(\mathbf{1}_n, \mathbf{X})$ satisfies Assumption 6 with

$$\frac{1}{n}\mathbf{X}'(\mathbf{I}_n - \mathbf{P}_{\mathbf{1}_n})\mathbf{X} \to \mathbf{D}_0 \quad \text{as } n \to \infty.$$

To shorten the notations, we write in the following

$$\mathbf{y}_{ij} = \mathbf{y}_j - \mathbf{y}_i, \quad \mathbf{x}_{ij} = \mathbf{x}_j - \mathbf{x}_i, \quad \text{and} \quad \varepsilon_{ij} = \varepsilon_j - \varepsilon_i,$$

for $i, j = 1,...,n$. Note that

$$\mathbf{y}_{ij} = \beta' \mathbf{x}_{ij} + \varepsilon_{ij}$$

and μ cancels out of the formula.

Testing problem I. Consider the testing problem with the null hypothesis H_0 : $\beta = \mathbf{0}$; that is, $\mathbf{E}(\mathbf{U}(\mathbf{y}_i - \mathbf{y}_j)) = \mathbf{0}$ for all $i \neq j$. The test statistic one can use here is

$$n^{1/2}\text{vec}(\mathbf{AVE}\{\mathbf{U}_{ij}\mathbf{x}'_{ij}\}).$$

It is, under the null hypothesis, asymptotically equivalent to a multivariate rank test statistic

$$n^{-1/2}\text{vec}(\mathbf{R}'\hat{\mathbf{X}}),$$

where $\hat{\mathbf{X}} = (\mathbf{I}_n - \mathbf{P}_{\mathbf{1}_n})\mathbf{X}$. Recall that the matrix of spatial ranks $\mathbf{R}$ is obtained by transformations

$$\mathbf{y}_i \to \mathbf{R}_i = \mathbf{R}(\mathbf{y}_i), \; i = 1,...,n, \quad \text{and} \quad \mathbf{Y} \to \mathbf{R} = (\mathbf{R}_1,...,\mathbf{R}_n)'.$$

Under the null hypothesis

$$n^{-1/2}\text{vec}(\mathbf{R}'\hat{\mathbf{X}}) \to_d N_{pq}(\mathbf{0}, \mathbf{D}_0 \otimes \mathbf{B}),$$

where $\mathbf{B} = \mathbf{E}\{\mathbf{R}(\mathbf{y}_i)\mathbf{R}(\mathbf{y}_i)'\}$ and finally

$$Q^2 = Q^2(\mathbf{Y}) = n \cdot tr(\mathbf{R}'\mathbf{P}_{\hat{\mathbf{X}}}\mathbf{R}(\mathbf{R}'\mathbf{R})^{-1}) \to_d \chi^2_{pq}.$$

Unfortunately, this test is not affine invariant but an affine invariant test version can be found, for example, using the following natural inner standardization. Find a transformation matrix $\mathbf{S}^{-1/2}$ such that if we transform

$$\mathbf{y}_i \to \hat{\mathbf{R}}_i = \mathbf{R}(\mathbf{S}^{-1/2}\mathbf{y}_i) \quad \text{and} \quad \mathbf{Y} \to \hat{\mathbf{R}} = (\hat{\mathbf{R}}_1,...,\hat{\mathbf{R}}_n)'$$

then

$$p \cdot \hat{\mathbf{R}}'\hat{\mathbf{R}} = tr(\hat{\mathbf{R}}'\hat{\mathbf{R}})\mathbf{I}_p.$$

The transformation is then a Tyler-type transformation but using ranks instead of signs, and the test statistic is

$$Q^2 = Q^2(\mathbf{X},\mathbf{Y}) = \frac{np}{tr(\hat{\mathbf{R}}'\hat{\mathbf{R}})} \cdot |\mathbf{P}_{\hat{\mathbf{X}}}\hat{\mathbf{R}}|^2.$$

The statistic Q^2 has the limiting χ^2_{pq} null distribution.

Estimation problem. The estimate $\hat{\beta}$ corresponding the rank test minimizes

$$D_n(\beta) = \mathbf{AVE}\{|\mathbf{y}_{ij} - \beta'\mathbf{x}_{ij}| - |\mathbf{y}_{ij}|\}$$

or solves

$$\mathbf{AVE}\{\mathbf{U}_{ij}(\beta)\mathbf{x}'_{ij}\} = \mathbf{0},$$

where

$$\mathbf{U}_{ij}(\beta) = \mathbf{U}(\mathbf{y}_{ij} - \beta'\mathbf{x}_{ij}), \quad i,j = 1,...,n.$$

The solution $\hat{\beta}$ may be found as in the regular LAD regression but replacing observations and explaining variables by differences of observations and explaining variables, respectively. The algorithm then uses the two iteration steps:

1.
$$\mathbf{e}_{ij} \leftarrow \mathbf{y}_{ij} - \beta'\mathbf{x}_{ij}, \quad i \neq j.$$

2.
$$\beta \leftarrow \beta + \left[\mathbf{AVE}\{|\mathbf{e}_{ij}|^{-1}\mathbf{x}_{ij}\mathbf{x}_{ij}'\}\right]^{-1} \mathbf{AVE}\{\mathbf{x}_{ij}\mathbf{U}(\mathbf{e}_{ij})'\}.$$

The estimate $\hat{\beta}$ can be called a rank-based estimate as the estimating equation can also be written in the form

$$\mathbf{R}(\hat{\beta})'\mathbf{X} = \mathbf{0}$$

with

$$\mathbf{R}(\beta) = (\mathbf{R}_1(\beta), ..., \mathbf{R}_n(\beta))',$$

where $\mathbf{R}_i(\beta)$ is the spatial rank of $\mathbf{y}_i - \beta'\mathbf{x}_i$ among $\mathbf{y}_1 - \beta'\mathbf{x}_1, ..., \mathbf{y}_n - \beta'\mathbf{x}_n$. See also Zhou (2009) for this estimate and its properties.

As before, it is possible to show that

$$\sqrt{n}\,\mathrm{vec}((\hat{\beta} - \beta)') \to_d N_{qp}\left(\mathbf{0}, \mathbf{D}_0^{-1} \otimes (\mathbf{A}^{-1}\mathbf{B}\mathbf{A}^{-1})\right),$$

where

$$\mathbf{A} = \mathbf{E}\left\{\mathbf{A}(\varepsilon_i - \varepsilon_j)\right\} \quad \text{and} \quad \mathbf{B} = \mathbf{E}\left\{\mathbf{B}(\varepsilon_i - \varepsilon_j, \varepsilon_i - \varepsilon_k)\right\}$$

with distinct i, j, and k, and

$$\mathbf{A}(\mathbf{y}) = \frac{1}{|\mathbf{y}|}\left[\mathbf{I}_p - \mathbf{U}(\mathbf{y})\mathbf{U}(\mathbf{y})'\right] \quad \text{and} \quad \mathbf{B}(\mathbf{y}_1, \mathbf{y}_2) = \mathbf{U}(\mathbf{y}_1)\mathbf{U}(\mathbf{y}_2)'.$$

Natural consistent estimates of $\mathbf{A}$ and $\mathbf{B}$ are then

$$\hat{\mathbf{A}} = \mathbf{AVE}\left\{\mathbf{A}\left(\mathbf{y}_{ij} - \hat{\beta}'\mathbf{x}_{ij}\right)\right\} \quad \text{and} \quad \hat{\mathbf{B}} = \mathbf{AVE}\left\{\mathbf{B}\left(\mathbf{y}_{ij} - \hat{\beta}'\mathbf{x}_{ij}, \mathbf{y}_{ik} - \hat{\beta}'\mathbf{x}_{ik}\right)\right\},$$

respectively. In fact, $\hat{\mathbf{B}}$ is simply the spatial rank covariance matrix of the estimated residuals.

As in the case of the regular LAD estimate, the estimate $\hat{\beta} = \hat{\beta}(\mathbf{X}, \mathbf{Y})$ is regression equivariant and $\mathbf{X}$ equivariant but not $\mathbf{Y}$ equivariant. The transformation retransformation estimation procedure can be created, for example, by first updating the residuals, then the β matrix, and finally the residual scatter matrix $\mathbf{S}$ as follows.

1.
$$\mathbf{e}_{ij} \leftarrow \mathbf{S}^{-1/2}(\mathbf{y}_{ij} - \beta'\mathbf{x}_{ij}), \quad i, j = 1, ..., n.$$

2.
$$\beta \leftarrow \beta + \left[\mathbf{AVE}\{|\mathbf{e}_{ij}|^{-1}\mathbf{x}_{ij}\mathbf{x}_{ij}'\}\right]^{-1} \mathbf{AVE}\{\mathbf{x}_{ij}\mathbf{U}(\mathbf{e}_{ij})'\}\mathbf{S}^{1/2}.$$

3.
$$\mathbf{S} \leftarrow p\, \mathbf{S}^{1/2}\, \mathbf{AVE}\{\mathbf{U}(\mathbf{e}_{ij})\mathbf{U}(\mathbf{e}_{ik})'\}\, \mathbf{S}^{1/2}.$$

Testing problem II. Consider again the model with two parts of explaining variables, $\mathbf{X}_1$ and $\mathbf{X}_2$:

$$\mathbf{Y} = \mathbf{1}_n \mu' + \mathbf{X}_1 \beta_1 + \mathbf{X}_2 \beta_2 + \varepsilon.$$

We wish to test the null hypothesis $H_0: \beta_2 = 0$. In the null case, the estimate of β_1 solves

$$\mathbf{R}(\hat{\beta}_1, \mathbf{0})' \mathbf{X}_1 = \mathbf{0}.$$

Then write

$$\hat{\mathbf{R}} = \mathbf{R}(\hat{\beta}_1, \mathbf{0}) \quad \text{and} \quad \hat{\mathbf{X}}_2 = (\mathbf{I}_n - \mathbf{P}_{\mathbf{X}_1})\mathbf{X}_2.$$

Then $\hat{\mathbf{R}}'\mathbf{X}_1 = \mathbf{0}$ and $\hat{\mathbf{X}}_2'\mathbf{X}_1 = \mathbf{0}$. The test statistic for testing $H_0: \beta_2 = 0$ is then

$$Q^2 = n \cdot tr\left(\hat{\mathbf{R}}'\mathbf{P}_{\hat{\mathbf{X}}_2}\hat{\mathbf{R}}(\hat{\mathbf{R}}'\hat{\mathbf{R}})^{-1}\right)$$

with a limiting chi-square distribution with $q_2 p$ degrees of freedom.

Unfortunately, the test statistic Q^2 is not invariant under affine transformations to $\mathbf{Y}$. The affine invariant version of the test is obtained if the spatial sign scores are obtained by using both inner centering and inner standardization. Then the spatial signs

$$\hat{\mathbf{R}}_i = \mathbf{R}\left(\mathbf{S}^{-1/2}(\mathbf{y}_i - \hat{\beta}_1 \mathbf{x}_{1i})\right), \quad i = 1, \dots, n,$$

satisfy

$$\hat{\mathbf{R}}'\mathbf{X}_1 = \mathbf{0} \quad \text{and} \quad p \cdot \hat{\mathbf{R}}'\hat{\mathbf{R}} = tr(\hat{\mathbf{R}}'\hat{\mathbf{R}})\mathbf{I}_p.$$

The test statistic is then

$$Q^2 = \frac{np}{tr(\hat{\mathbf{R}}'\hat{\mathbf{R}})} \cdot |\mathbf{P}_{\hat{\mathbf{X}}_2}\hat{\mathbf{R}}|^2.$$

13.5 An example

The dataset considered in this example is the LASERI data already analyzed in Chapter 10. We consider the multivariate regression problem where the response variables are the differences HRT1T2, COT1T2, and SVRIT1T2, and the explaining variables are sex (0/1), age (years), and WHR (waist to hip ratio). See Figure 13.1 for the scatterplot matrix. The variables HRT1T2, COT1T2, and SVRIT1T2 measure the reaction of the individual hemodynamic system to the change in positions.

We first estimate the regression coefficient matrix in the full model with three explaining variables: sex, age, and WHR. If the spatial sign score (LAD) with inner standardization is used, one gets

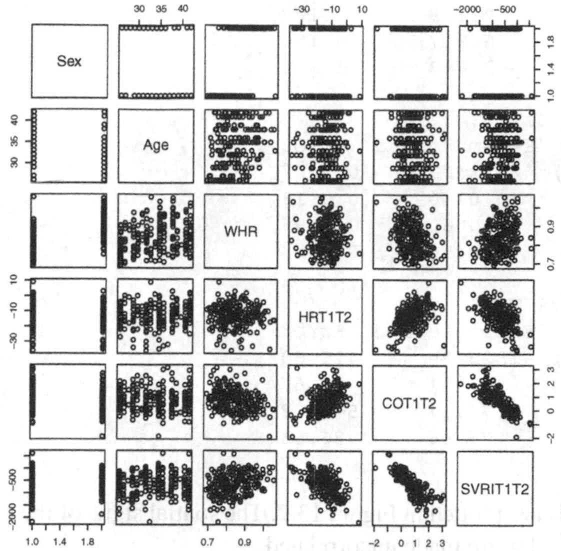

Fig. 13.1 Pairwise scatterplots for the variables used in the regression analysis.

```
> data(LASERI)
> with(LASERI, pairs( cbind(Sex, Age, WHR, HRT1T2,
      COT1T2, SVRIT1T2 )))
>
> is.reg.fullmodel <- mv.l1lm(cbind(HRT1T2, COT1T2, SVRIT1T2)
  ~ Age +  WHR + Sex, data=LASERI, score="s", stand="i")
> with(LASERI, pairs( cbind(Sex, Age, WHR,
  residuals(is.reg.fullmodel))))
> summary(is.reg.fullmodel)

Multivariate regression using spatial sign scores
and inner standardization

Call:
mv.l1lm(formula = cbind(HRT1T2, COT1T2, SVRIT1T2) ~
Age + WHR +  Sex, scores = "s", stand = "i", data = LASERI)

Testing that all coefficients = 0:
Q.2 = 213.0913 with 12 df, p.value < 2.2e-16

Results by response:

Response HRT1T2 :
              Estimate Std. Error
(Intercept)   -21.713       6.249
```

```
Age              0.169        0.100
WHR              5.337        8.154
SexMale         -2.648        1.300

Response COT1T2 :
              Estimate Std. Error
(Intercept)    2.3401     0.7047
Age            0.0140     0.0113
WHR           -2.5060     0.9196
SexMale       -0.0496     0.1467

Response SVRIT1T2 :
              Estimate Std. Error
(Intercept) -1525.80     415.80
Age            -3.31       6.67
WHR          1173.87     542.58
SexMale        76.06      86.53
```

The residuals are plotted in Figure 13.2. The spatial signs of the residuals and the explaining variables are made uncorrelated.

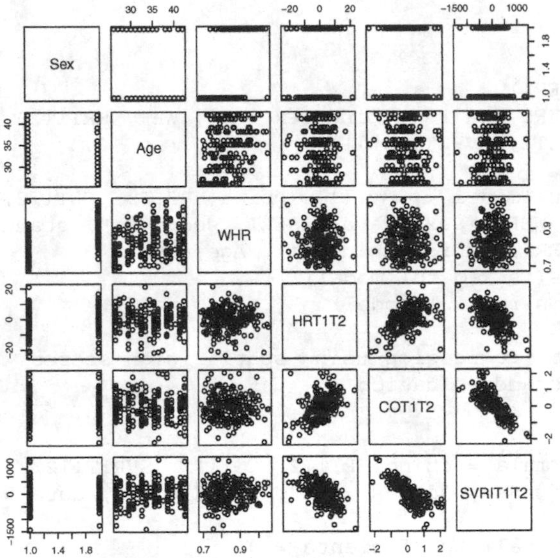

Fig. 13.2 Residual plots for the estimated full model with the spatial sign score for LASERI data.

If one uses the identity score instead, regular L_2 analysis gives quite similar results. See the results below.

```
> is.reg.fullmodel2 <- mv.l1lm(cbind(HRT1T2, COT1T2, SVRIT1T2)
  ~ Age +  WHR + Sex, data=LASERI)
> summary(is.reg.fullmodel2)

Multivariate regression using identity scores

Call:
mv.l1lm(formula = cbind(HRT1T2, COT1T2, SVRIT1T2)
 ~ Age + WHR +     Sex, data = LASERI)

Testing that all coefficients = 0:
Q.2 = 225.7625 with 12 df, p.value < 2.2e-16

Results by response:

Response HRT1T2 :
            Estimate Std. Error
(Intercept)  -21.013      6.282
Age            0.140      0.101
WHR            5.146      8.197
SexMale       -2.510      1.307

Response COT1T2 :
            Estimate Std. Error
(Intercept)   3.0223     0.6406
Age           0.0105     0.0103
WHR          -3.1486     0.8359
SexMale      -0.0084     0.1333

Response SVRIT1T2 :
            Estimate Std. Error
(Intercept) -1834.03     365.30
Age            -1.74       5.86
WHR          1462.88     476.69
SexMale        63.27      76.02
```

If one wishes to test the hypothesis that the variable WHR has no effect on the response variables, one can first estimate the parameters in the submodel (without WHR) and then use the score test as described earlier. One then gets

```
> is.reg.submodel <- mv.l1lm(cbind(HRT1T2, COT1T2, SVRIT1T2)
  ~ Age + Sex, data=LASERI, score="s", stand="i")
> anova(is.reg.fullmodel,is.reg.submodel)

Comparisons between multivariate linear models
```

```
Full model:          mv.l1lm(formula =
cbind(HRT1T2, COT1T2, SVRIT1T2)
~ Age + WHR +        Sex, scores = "s", stand = "i", data = LASERI)
Restricted model: mv.l1lm(formula =
cbind(HRT1T2, COT1T2, SVRIT1T2)
~ Age + Sex,        scores = "s", stand = "i", data = LASERI)

Score type test that coefficients not
in the restricted model are 0:
Q.2 = 18.7278 with 3 df, p.value = 0.0003112
```

13.6 Other approaches

Rao (1988) proposed the use of univariate LAD regression separately for the p response variable. Puri and Sen (1985), Section 6.4, and Davis and McKean (1993) developed multivariate regression methods based on coordinatewise ranks. Chakraborty (1999) used the transformation retransformation technique with marginal LAD estimates to find affine equivariant versions of the LAD estimates.

Multivariate spatial sign methods have been studied in Bai et al. (1990) and Arcones (1998). Multivariate affine equivariant regression quantiles based on spatial signs and the transformation retransformation technique were introduced and discussed in Chakraborty (2003). Asymptotics for the spatial rank methods were considered in Zhou (2009).

Theil-type estimates based on the Oja median were given in Busarova et al. (2006) and Shen (2009). For a different type of regression coefficient estimates that are based on the Oja sign and rank covariance matrices, see Ollila et al. (2002, 2004b).

Chapter 14
Analysis of cluster-correlated data

Abstract In this chapter it is shown how the spatial sign and rank methods can be extended to cluster-correlated data. Tests and estimates for the one-sample location problem with a general score function are given in detail. Then two-sample weighted spatial rank tests are considered.

14.1 Introduction

In previous chapters we assumed that the observations in $\mathbf{Y} = (\mathbf{y}_1, ..., \mathbf{y}_n)'$ are generated by the model

$$\mathbf{Y} = \mathbf{X}\beta + \varepsilon,$$

where the n p-variate residuals, that is, the rows of $\varepsilon = (\varepsilon_1, ..., \varepsilon_n)'$, are independent and identically distributed (i.i.d.) random vectors. The assumption that the observations are independent is not true, however, if the data are clustered.

Clustered data can arise in a variety of applications. There may occur natural groups in the target population. These groupings may, for example, be based on clinics for patients, schools for students, litters for rats, and so on. Still one example on clustered data is the data arising in longitudinal studies. Then the measurements on the individuals (clusters) are taken repeatedly over a time interval. If the clustering in the data is simply ignored, in some cases there can be a serious underestimation of the variability of the estimators. The true standard deviation of the sample mean as an estimator of the population mean, for example, may be much larger than its estimate under the i.i.d. assumption. This underestimation will further result in confidence intervals that are too narrow and p-values that are too small. Therefore an adjustment to standard statistical methods depending on cluster sizes and intraclass correlation is needed.

Traditionally, parametric mixed models have been used to account for the correlation structures among the dependent observational units. Then one assumes that

H. Oja, *Multivariate Nonparametric Methods with R: An Approach Based on Spatial Signs and Ranks*, Lecture Notes in Statistics 199, DOI 10.1007/978-1-4419-0468-3_14,

$$\mathbf{Y} = \mathbf{Z}\alpha + \mathbf{X}\beta + \varepsilon$$

where $\mathbf{Z}$ is an $n \times d$ matrix of group (cluster) membership. (The jth column indicates the group membership in the jth cluster.) The d rows of the $d \times p$ matrix α give the random effect of d clusters. The idea is that the clusters in the sample are not fixed but a random sample from a population of clusters. Often the data are collected in clusters, and the approximate distributions are obtained letting $d \to \infty$. In the mixed model approach the rows of α and ε are assumed to be independent and multinormally distributed. One can again relax the assumptions and use multivariate spatial sign and rank scores in the analysis. It also seems clear that the observations should have different weights in the analysis. The observations in a cluster of ten observations should not have the same weight inasmuch as a single observation in a cluster of size one as the information in the first cluster is not tenfold.

14.2 One-sample case

14.2.1 Notation and assumptions

Let

$$\mathbf{Y} = (\mathbf{y}_1, \mathbf{y}_2, ..., \mathbf{y}_n)'$$

be a sample of p-variate random vectors with sample size n. We assume now that the observations come in d clusters and that the $n \times d$ matrix

$$\mathbf{Z} = (\mathbf{z}_1, \mathbf{z}_2, ..., \mathbf{z}_n)'$$

gives the cluster membership so that

$$z_{ij} = \begin{cases} 1, & \text{if the } i\text{th observation comes from cluster } j \\ 0, & \text{otherwise.} \end{cases}$$

Note that

$$(\mathbf{Z}\mathbf{Z}')_{ij} = \begin{cases} 1, & \text{if the } i\text{th and } j\text{th observations come from the same cluster} \\ 0, & \text{otherwise,} \end{cases}$$

and that $\mathbf{Z}'\mathbf{Z}$ is a $d \times d$ diagonal matrix whose diagonal elements are the d cluster sizes, say $m_1, ..., m_d$.

The one-sample parametric location model with random cluster effects is often written as

$$\mathbf{Y} = \mathbf{Z}\alpha + \mathbf{1}_n\mu' + \varepsilon,$$

where the d rows of α are i.i.d. from $N_p(\mathbf{0}, \Omega)$ and the n rows of ε are i.i.d. from $N_p(\mathbf{0}, \Sigma)$, and α and ε are independent. The model can be reformulated as

$$\mathbf{Y} = \mathbf{1}_n\mu' + \varepsilon,$$

where now

$$\text{vec}(\varepsilon') \sim N_{np}(\mathbf{0}, \mathbf{I}_n \otimes \Sigma + \mathbf{ZZ}' \otimes \Omega).$$

We thus move the cluster effect to the covariance matrix of the error variable. If $\varepsilon = (\varepsilon_1, \ldots, \varepsilon_n)'$ then the model states that

1. $\varepsilon_i \sim N_p(\mathbf{0}, \Sigma + \Omega)$ for all $i = 1, \ldots, n$.
2. If $(\mathbf{ZZ}')_{ij} = 1$, $i \neq j$, then

$$\text{vec}(\varepsilon_i, \varepsilon_j) \sim N_{2p}\left(\mathbf{0}, \begin{pmatrix} \Sigma + \Omega & \Omega \\ \Omega & \Sigma + \Omega \end{pmatrix}\right).$$

3. If $(\mathbf{ZZ}')_{ij} = 0$ then ε_i and ε_j are independent.

In our approach, we relax the assumptions. We still assume that

$$\mathbf{Y} = \mathbf{1}_n\mu' + \varepsilon$$

but the residuals $\varepsilon_1, ..., \varepsilon_n$ now satisfy the following.

Assumption 7 *The rows of $\varepsilon = (\varepsilon_1, ..., \varepsilon_n)'$ satisfy*

1. $\varepsilon_i \sim -\varepsilon_i$ and $\varepsilon_i \sim \varepsilon_j$, for all $i, j = 1, ..., n$.
2. $(\varepsilon_i, \varepsilon_j) \sim (\varepsilon_{i'}, \varepsilon_{j'})$ for all $i \neq j$ and $i' \neq j'$ and $(\mathbf{ZZ}')_{ij} = (\mathbf{ZZ}')_{i'j'}$.
3. If $(\mathbf{ZZ}')_{ij} = 0$ then ε_i and ε_j are independent.

14.2.2 Tests and estimates

A general idea to construct tests and estimates is again to use an odd vector-valued score function $\mathbf{T}(\mathbf{y})$ to calculate individual scores $\mathbf{T}_i = \mathbf{T}(\mathbf{y}_i)$, $i = 1, ..., n$. We write

$$\mathbf{T} = (\mathbf{T}_1, \mathbf{T}_2, ..., \mathbf{T}_n)'.$$

We need the assumption that $\mathbf{E}(|\mathbf{T}(\varepsilon_i)|^{2+\gamma})$ is bounded for some $\gamma > 0$. Let $\mathbf{L}(\mathbf{y})$ be the optimal location score function, that is, the gradient vector of $\log f(\mathbf{y} - \mu)$ with respect to μ at the origin. Here $f(\mathbf{y})$ is the density of ε_i. If $H_0 : \mu = \mathbf{0}$ is true, then $E(\mathbf{T}_i) = 0$ for all $i = 1, ..., n$. Write as before

$$\mathbf{A} = \mathbf{E}(\mathbf{T}(\varepsilon_i)\mathbf{L}(\varepsilon_i)') \quad \text{and} \quad \mathbf{B} = \mathbf{E}(\mathbf{T}(\varepsilon_i)\mathbf{T}(\varepsilon_i)').$$

We now also need covariances of two distinct transformed residuals in the same cluster; that is,

$$\mathbf{C} = \mathbf{E}(\mathbf{T}(\varepsilon_i)\mathbf{T}(\varepsilon_j)') \quad \text{where } i \neq j \text{ satisfy } (\mathbf{ZZ}')_{ij} = 1.$$

Clearly

$$\mathbf{COV}(\mathrm{vec}(\mathbf{T}')) = \mathbf{I}_n \otimes \mathbf{B} + (\mathbf{ZZ}' - \mathbf{I}_n) \otimes \mathbf{C}$$

For the sampling design, we have the next assumption.

Assumption 8 *Assume that*

$$\frac{1}{n}\mathbf{1}_n'(\mathbf{ZZ}' - \mathbf{I}_n)\mathbf{1}_n = \frac{1}{n}\sum_{i=1}^{d} m_i^2 - 1 \to d_o$$

as $d \to \infty$.

Then, for the one-sample location problem,

- The test for $H_0: \ \mu = \mathbf{0}$ is based on

$$\mathbf{AVE}\{\mathbf{T}(\mathbf{y}_i)\} = \frac{1}{n}(\mathbf{1}_n' \otimes \mathbf{I}_p)\mathrm{vec}(\mathbf{T}').$$

- The companion location estimate $\hat{\mu}$ is determined by the estimating equation

$$\mathbf{AVE}\{\mathbf{T}(\mathbf{y}_i - \hat{\mu})\} = \mathbf{0}.$$

Consider the null hypothesis $H_0: \ \mu = \mathbf{0}$ and a (contiguous) sequence of alternatives $H_n: \ \mu = n^{-1/2}\delta$. Then (under general assumptions that should be proved separately for each choice of the score function)

- Under the null hypothesis H_0,

$$\sqrt{n}\,\mathbf{AVE}\{\mathbf{T}(\mathbf{y}_i)\} \to_d N_p(\mathbf{0}, \mathbf{B} + d_0\mathbf{D}).$$

- Under the sequence of alternatives H_n,

$$Q^2 = \mathbf{1}_n'\mathbf{T}(\mathbf{T}'\mathbf{ZZ}'\mathbf{T})^{-1}\mathbf{T}'\mathbf{1}_n \to_d \chi_p^2(\delta'\mathbf{A}(\mathbf{B} + d_0\mathbf{C})^{-1}\mathbf{A}\delta).$$

- $\sqrt{n}(\hat{\mu} - \mu) \to_d N_p(\mathbf{0}, \mathbf{A}^{-1}(\mathbf{B} + d_0\mathbf{C})\mathbf{A}^{-1})$.

Note that the limiting covariance matrix of $\hat{\mu}$ needs an adjustment

$$\mathbf{A}^{-1}\mathbf{B}\mathbf{A}^{-1} \ \to \ \mathbf{A}^{-1}(\mathbf{B} + d_0\mathbf{C})\mathbf{A}^{-1},$$

and the estimated confidence ellipsoid without this correction is too small (positive intracluster correlation).

14.2.3 Tests and estimates, weighted versions

In this section we consider the tests and estimates based on the weighted scores. Let

$$\mathbf{W} = diag(w_1,...,w_n)$$

be an $n \times n$ diagonal matrix with a non-negative weight w_i associated with the ith observation. The weight matrix is assumed to be fixed and possibly defined by the cluster structure $\mathbf{Z}$. The covariance structure of the weighted score matrix $\mathbf{WT}$ is then

$$\mathbf{COV}\left(\mathrm{vec}(\mathbf{T}'\mathbf{W})\right) = \mathbf{W}^2 \otimes \mathbf{B} + \left(\mathbf{W}(\mathbf{ZZ}' - \mathbf{I}_n)\mathbf{W}\right) \otimes \mathbf{C}.$$

For the cluster structure and the weights, we now need the following assumption.

Assumption 9 *There exist constants d_1 and d_2 such that*

$$\frac{1}{n}\mathbf{1}_n'\mathbf{W}^2\mathbf{1}_n \to d_1 \quad \text{and} \quad \frac{1}{n}\mathbf{1}_n'\mathbf{W}\left(\mathbf{ZZ}' - \mathbf{I}_n\right)\mathbf{W}\mathbf{1}_n \to d_2$$

as d tends to infinity.

Then, for the one-sample location problem,

- The test is based on

$$\mathbf{AVE}\left\{w_i\mathbf{T}(\mathbf{y}_i)\right\} = \frac{1}{n}\mathbf{T}'\mathbf{W}\mathbf{1}_n = \frac{1}{n}(\mathbf{1}_n' \otimes \mathbf{I}_p)\mathrm{vec}(\mathbf{T}'\mathbf{W}).$$

- The companion location estimate $\hat{\mu}$ is determined by the estimating equation

$$\mathbf{AVE}\{w_i\mathbf{T}(\mathbf{y}_i - \hat{\mu})\} = \mathbf{0}.$$

Consider the null hypothesis $H_0: \ \mu = \mathbf{0}$ and a (contiguous) sequence of alternatives $H_n: \ \mu = n^{-1/2}\delta$. Then (again under certain assumptions depending on the score function),

- Under the null hypothesis H_0,

$$\sqrt{n}\,\mathbf{AVE}\left\{w_i\mathbf{T}(\mathbf{x}_i)\right\} \to_d N_p(\mathbf{0}, d_1\mathbf{B} + d_2\mathbf{C}).$$

- Under the sequence of alternatives H_n,

$$Q^2 = \mathbf{1}_n'\mathbf{WT}(\mathbf{T}'\mathbf{WZZ}'\mathbf{WT})^{-1}\mathbf{T}'\mathbf{W}\mathbf{1}_n \to_d \chi_p^2(\delta'\mathbf{A}(d_1\mathbf{B} + d_2\mathbf{C})^{-1}\mathbf{A}\delta).$$

- $\sqrt{n}(\hat{\mu} - \mu) \to_d N_p(\mathbf{0}, \mathbf{A}^{-1}(d_1\mathbf{B} + d_2\mathbf{C})\mathbf{A}^{-1})$.

How should one then choose the weights? Using the results above, one can choose the weights to maximize Pitman efficiency of the test or to minimize the determinant of the covariance matrix of the estimate, for example. Explicit solutions can be found in some simplified cases. If $\mathbf{C} = \rho\mathbf{B}$ (ρ is the intraclass correlation) then the covariance matrix has the structure

$$\mathbf{COV}(\mathrm{vec}(\mathbf{T}')) = \Sigma \otimes \mathbf{B}, \qquad \text{where } \Sigma = \mathbf{I}_n + \rho(\mathbf{ZZ}' - \mathbf{I}_n).$$

One can then use the Lagrange multiplier technique to find the optimal weights $\mathbf{w} = (w_1, ..., w_n)'$. The solution is

$$\mathbf{w} = \lambda \Sigma^{-1} \mathbf{1}_n,$$

where λ is the Lagrange multiplier chosen so that the constraint $\mathbf{w}'\mathbf{1}_n = n$ is satisfied. The weights in the ith cluster are then proportional to $[1+(m_i-1)\rho]^{-1}$ (Larocque et al. (2007)). The larger the cluster size, the smaller are the weights.

We end this section with the notion that the proposed score-based testing and estimation procedures are not necessarily affine invariant and equivariant, respectively. Again, affine invariant and equivariant versions may be obtained, as before, using the transformation retransformation technique. Natural unweighted and weighted scatter matrix estimates for this purpose have not yet been developed.

14.3 Two samples: Weighted spatial rank test

Assume that $(\mathbf{X}, \mathbf{Z}, \mathbf{Y})$ is the data matrix and consider first the general linear regression model

$$\mathbf{Y} = \mathbf{X}\beta + \varepsilon,$$

where as before $\mathbf{Y} = (\mathbf{y}_1, ..., \mathbf{y}_n)'$ is an $n \times p$ matrix of n observed values of p response variables, $\mathbf{X} = (\mathbf{x}_1, ..., \mathbf{x}_n)'$ is an $n \times q$ matrix of observed values of q explaining variables, β is a $q \times p$ matrix of regression coefficients, and $\varepsilon = (\varepsilon_1, ..., \varepsilon_n)'$ is an $n \times p$ matrix of residuals. At the individual level, one can then write

$$\mathbf{y}_i = \beta'\mathbf{x}_i + \varepsilon_i, \quad i = 1, ..., n.$$

As before, the matrix $\mathbf{Z}$ is the $n \times d$ matrix indicating the cluster membership. The residuals $\varepsilon_1, ..., \varepsilon_n$ are not iid any more but satisfy

Assumption 10 *The rows of* $\varepsilon = (\varepsilon_1, ..., \varepsilon_n)'$ *satisfy*

1. $\mathbf{E}(\mathbf{U}(\varepsilon_i)) = \mathbf{0}$ *and* $\varepsilon_i \sim \varepsilon_j$, *for all* $i, j = 1, ..., n$.
2. $(\varepsilon_i, \varepsilon_j) \sim (\varepsilon_{i'}, \varepsilon_{j'})$ *for all* $i \neq j$ *and* $i' \neq j'$ *and* $(\mathbf{ZZ}')_{ij} = (\mathbf{ZZ}')_{i'j'}$.
3. *If* $(\mathbf{ZZ}')_{ij} = 0$ *then* ε_i *and* ε_j *are independent.*

The first condition says that all the p-variate distributions of the ε_i are the same; no symmetry condition is needed. The condition $\mathbf{E}(\mathbf{U}(\varepsilon_i)) = \mathbf{0}$ is used here just to fix the center of the distribution of the residuals. (The spatial median of the residuals is zero.)

Next we focus on the two-sample location model where

$$\mathbf{X} = (\mathbf{1}_n, \mathbf{x}) \quad \text{and} \quad \beta = (\mu, \Delta)',$$

the n-vector $\mathbf{x}$ is the indicator for the second sample membership, and μ and $\mu + \Delta$ are the two location centers (spatial medians of the two populations). We wish to test the null hypothesis $H_0 : \Delta = \mathbf{0}$ and estimate the value of unknown Δ.

Note that the sample sizes are then

$$n_1 = \mathbf{1}_n'(\mathbf{1}_n - \mathbf{x}) \quad \text{and} \quad n_2 = \mathbf{1}_n'\mathbf{x},$$

and the cluster sizes $m_1, ..., m_d$ are the diagonal elements of the $d \times d$ diagonal matrix $\mathbf{Z}'\mathbf{Z}$. The sample design is given by the frequency table for group and cluster membership, that is,

$$\left((\mathbf{1}_n - \mathbf{x})'\mathbf{Z}, \mathbf{x}'\mathbf{Z}\right).$$

If the null hypothesis $H_0 : \Delta = \mathbf{0}$ is true then the observations $\mathbf{y}_1, ..., \mathbf{y}_n$ are i.i.d. from a distribution with the cdf F, say. The population spatial rank score function is then

$$\mathbf{R}_F(\mathbf{y}) = \mathbf{E}(\mathbf{U}(\mathbf{y} - \mathbf{y}_i)).$$

Function $\mathbf{R}_F$ is naturally unknown. An often improved estimate of the population spatial rank function is obtained if one uses a weighted spatial rank function

$$\mathbf{R}_w(\mathbf{y}) = \mathbf{AVE}\{w_i \mathbf{U}(\mathbf{y} - \mathbf{y}_i)\}$$

with some positive strategically chosen individual weights $w_1, ..., w_n$. We again write

$$\mathbf{w} = (w_1, ..., w_n)' \quad \text{and} \quad \mathbf{W} = diag(w_1, ..., w_n).$$

Also write in short

$$\mathbf{R}_F = (\mathbf{R}_F(\mathbf{y}_1), ..., \mathbf{R}_F(\mathbf{y}_n)) \quad \text{and} \quad \mathbf{R}_w = (\mathbf{R}_w(\mathbf{y}_1), ..., \mathbf{R}_w(\mathbf{y}_n)).$$

Note that the weighted ranks $\mathbf{R}_w(\mathbf{y}_1), ..., \mathbf{R}_w(\mathbf{y}_n)$ are now centered in the sense that

$$\mathbf{R}_w'\mathbf{W}\mathbf{1}_n = \mathbf{R}_w'\mathbf{w} = 0.$$

The test statistic is then based on the weighted sum of weighted ranks over the second sample, that is,

$$\mathbf{R}_w'\mathbf{W}\mathbf{X}.$$

One can then show that, under the null hypothesis and under some general assumptions,

$$\frac{1}{\sqrt{n}}\mathbf{R}_w'\mathbf{W}\mathbf{x} = \frac{1}{\sqrt{n}}\mathbf{R}_F'\mathbf{W}\mathbf{x}_w + o_P(1),$$

where

$$\mathbf{x}_w = \left(\mathbf{I}_n - \frac{1}{n}\mathbf{1}_n\mathbf{1}_n'\mathbf{W}\right)\mathbf{x}.$$

Note that now the $\mathbf{x}_w$ are centered (instead of ranks) so that $\mathbf{x}_w'\mathbf{w} = 0$. Thus the limiting null distribution of $n^{-1/2}\mathbf{R}_w'\mathbf{W}\mathbf{x}$ is a p-variate normal distribution with mean value zero and covariance matrix $d_1\mathbf{B} + d_2\mathbf{C}$ where

$$\mathbf{B} = \mathbf{E}(\mathbf{R}_F(\varepsilon_i)\mathbf{R}_F(\varepsilon_i)')$$

and

$$\mathbf{C} = \mathbf{E}(\mathbf{R}_F(\varepsilon_i)\mathbf{R}_F(\varepsilon_j)') \quad \text{with } i \neq j \text{ such that } (\mathbf{ZZ}')_{ij} = 1.$$

Here it is assumed that

$$\mathbf{x}_w'\mathbf{W}^2\mathbf{x}_w \to d_1 \quad \text{and} \quad \mathbf{x}_w'\mathbf{W}(\mathbf{ZZ}' - \mathbf{I}_n)\mathbf{W}\mathbf{x}_w \to d_2.$$

Finally, if $\mathbf{G}$ is a diagonal matrix with diagonal elements given in $2\mathbf{x} - \mathbf{1}_n$ then the limiting null distribution of the squared version

$$Q^2 = \mathbf{1}_n'\mathbf{G}\mathbf{W}\mathbf{R}_w\left(\mathbf{R}_w'\mathbf{G}\mathbf{W}\mathbf{Z}\mathbf{Z}'\mathbf{W}\mathbf{G}\mathbf{R}_w\right)^{-1}\mathbf{R}_w\mathbf{W}\mathbf{G}\mathbf{1}_n$$

is a chi-squared distribution with p degrees of freedom. This extends the Wilcoxon-Mann-Whitney test to the case of multivariate clustered data. See Nevalainen et al. (2009) for more details and for the several-sample case.

14.4 References and other approaches

Most theoretical work for the analysis of longitudinal or clustered data concerns univariate continuous response variables having a normal distribution. Rosner and Grove (1999) and Rosner et al. (2003) generalized the standard Wilcoxon-Mann-Whitney rank sum test to the cluster-correlated case with cluster members belonging to the same treatment groups. Datta and Satten (2005) and Datta and Satten (2008) developed the rank-sum tests for cases where members in the same cluster may belong to different treatment groups. Additionally, the correlation between cluster members may depend on the cluster size. Finally, Rosner et al. (2003) derived an adjusted variance estimate for a randomization-based Wilcoxon signed rank test for clustered paired data. They also introduced a weighted signed-rank statistic to attain better efficiency. The weighted multivariate sign test is the only nonparametric multivariate test for cluster-correlated data considered in the literature thus far.

Quite recently, Larocque (2003), Nevalainen et al. (2007a,b, 2009), Larocque et al. (2007), and Haataja et al. (2008) have developed extensions of the sign and signed-rank tests and the corresponding estimates to multivariate cluster-correlated data. The results presented here are based on these papers.

Appendix A
Some vector and matrix algebra

An $r \times s$ *matrix* $\mathbf{A}$ is an array of real numbers

$$\mathbf{A} = \begin{pmatrix} a_{11} & a_{12} & \dots & a_{1s} \\ a_{21} & a_{22} & \dots & a_{2s} \\ \dots & \dots & \dots & \dots \\ a_{r1} & a_{r2} & \dots & a_{rs} \end{pmatrix}.$$

The number a_{ij} is called the (i,j) *element* of $\mathbf{A}$. The set of $r \times s$ matrices is here denoted by $\mathscr{M}(r,s)$. An $r \times s$ *zero matrix*, written $\mathbf{0}$, is a matrix with all elements zero and is the zero element in $\mathscr{M}(r,s)$.

$r \times 1$ matrices are called *(column) vectors* or *r-vectors*; $1 \times s$ matrices are *row vectors*. Column vectors are denoted by bold lower-case letters $\mathbf{a}, \mathbf{b}, \ldots$. A 1×1 matrix a is just a real number. A set of vectors $\mathbf{a}_1, \ldots, \mathbf{a}_r$ is said to be *linearly dependent* if there exist scalars $c_1, \ldots, c_r$, not all zero, such that $c_1\mathbf{a}_1 + \cdots + c_r\mathbf{a}_r = \mathbf{0}$. Otherwise they are *linearly independent*. Write $\mathbf{e}_i$, $i = 1, \ldots, r$ for an r-vector with ith element one and other elements zero. These vectors are linearly independent and give an orthonormal base for $\mathfrak{R}^r$.

The *transpose* of $\mathbf{A}$, written as $\mathbf{A}'$, is the $s \times r$ matrix

$$\mathbf{A} = \begin{pmatrix} a_{11} & a_{21} & \dots & a_{r1} \\ a_{12} & a_{22} & \dots & a_{r2} \\ \dots & \dots & \dots & \dots \\ a_{1s} & a_{2s} & \dots & a_{rs} \end{pmatrix},$$

obtained by interchanging the roles of the rows and columns; the ith row becomes the ith column and the jth column becomes the jth row, $i = 1, \ldots, r$ and $j = 1, \ldots, s$.

The *sum* of two $r \times s$ matrices $\mathbf{A}$ and $\mathbf{B}$ is again an $r \times s$ matrix $\mathbf{C} = \mathbf{A} + \mathbf{B}$, whose (i,j) element is

$$c_{ij} = a_{ij} + b_{ij}.$$

H. Oja, *Multivariate Nonparametric Methods with R: An Approach Based on Spatial Signs and Ranks*, Lecture Notes in Statistics 199, DOI 10.1007/978-1-4419-0468-3,

The *scalar product* of real number c and $r \times s$ matrix $\mathbf{A}$ is an $r \times s$ matrix with (i,j)-element $c \cdot a_{ij}$. The *product* of an $r \times s$ matrix $\mathbf{A}$ and an $s \times t$ matrix $\mathbf{B}$ is an $r \times t$ matrix $\mathbf{C} = \mathbf{AB}$, whose (i,j) element is

$$c_{ij} = \sum_{k=1}^{s} a_{ik} b_{kj}.$$

An $r \times r$ matrix $\mathbf{A}$ is a *square matrix* (the number of rows equals the number of columns). Square matrix $\mathbf{A}$ is *symmetric* if $\mathbf{A}' = \mathbf{A}$. Square matrix $\mathbf{A}$ is a *diagonal matrix* if $a_{ij} = 0$ whenever $i \neq j$ (off-diagonal elements are zero). $r \times r$ diagonal matrix

$$\mathbf{I}_r = \begin{pmatrix} 1 & 0 & \ldots & 0 \\ 0 & 1 & \ldots & 0 \\ \ldots & \ldots & \ldots & \ldots \\ 0 & 0 & \ldots & 1 \end{pmatrix}$$

is called an *identity matrix* (diagonal elements are one, off-diagonal elements zero). Note that

$$\mathbf{I}_r = \sum_{i=1}^{r} \mathbf{e}_i \mathbf{e}_i'.$$

If $\mathbf{A}$ is an $r \times s$ matrix then $\mathbf{I}_r \mathbf{A} = \mathbf{A}$ and $\mathbf{A}\mathbf{I}_s = \mathbf{A}$. For any r-vector $\mathbf{a}$, diag($\mathbf{a}$) is a diagonal matrix with r diagonal elements given by $\mathbf{a} = (a_1, \ldots, a_r)'$ (in the same order). On the other hand, we also write diag($\mathbf{A}$) for a diagonal matrix with the diagonal elements as in $\mathbf{A}$.

A *permutation matrix* $\mathbf{P}_r$ is obtained by permuting the rows and/or the columns of $\mathbf{I}_r$. The elements of the permutation matrix are then zeros and ones, and the row and column sums equal one. An elemental permutation

$$\mathbf{I}_r - \mathbf{e}_i \mathbf{e}_i' - \mathbf{e}_j \mathbf{e}_j' + \mathbf{e}_i \mathbf{e}_j' + \mathbf{e}_j \mathbf{e}_i'$$

is obtained just by interchanging the ith and jth row of the identity matrix. Permutation matrices can be given as a product of elemental permutations. Write $c(\mathbf{P}_r)$ for the smallest number of elemental permutations needed for transformation $\mathbf{I}_r \to \mathbf{P}_r$. The set of all $r \times r$ permutation matrices by $\mathscr{P}_r$ includes $r!$ different permutation matrices.

An $r \times r$ matrix $\mathbf{A}$ is called a *projection matrix* if it is *idempotent*, that is, if $\mathbf{A}^2 = \mathbf{A}$ and $\mathbf{A}' = \mathbf{A}$. If $\mathbf{A}$ is a projection matrix, then so is $\mathbf{I}_r - \mathbf{A}$.

$r \times r$ matrix $\mathbf{B}$ is the *inverse* of an $r \times r$ matrix $\mathbf{A}$ if

$$\mathbf{AB} = \mathbf{BA} = \mathbf{I}_r.$$

Then we write $\mathbf{B} = \mathbf{A}^{-1}$. A square matrix $\mathbf{A}$ is called *invertible* if its inverse $\mathbf{A}^{-1}$ exists. Clearly identity matrix $\mathbf{I}_r$ is invertible and $\mathbf{I}_r^{-1} = \mathbf{I}_r$. Permutation matrices are invertible with $\mathbf{P}_r^{-1} = \mathbf{P}_r'$. The inverse of a diagonal matrix $\mathbf{A} = \text{diag}(a_1, ..., a_r)$ exists if all diagonal elements are nonzero, and then $\mathbf{A}^{-1} = \text{diag}(a_1^{-1}, ..., a_r^{-1})$. Matrix $\mathbf{A}$ is called *orthogonal* if $\mathbf{A}^{-1} = \mathbf{A}'$. Note that permutation matrices are orthogonal.

The *determinant* of an $r \times r$ diagonal matrix $\mathbf{A}$ is

$$\det(\mathbf{A}) = a_{11}a_{22}\cdots a_{rr}.$$

More generally, the determinant of an $r \times r$ square matrix $\mathbf{A}$, written as $\det(A)$, is defined as

$$\det(\mathbf{A}) = \sum \left\{ (-1)^{c(\mathbf{P}_r)} \det(\text{diag}(\mathbf{P}_r \mathbf{A})) \right\}$$

where the sum is over all possible permutations $\mathbf{P}_r \in \mathscr{P}_r$.

The *trace* of an $r \times r$ square matrix $\mathbf{A}$, denoted by $tr(\mathbf{A})$, is the sum of diagonal elements, that is,

$$tr(\mathbf{A}) = a_{11} + a_{22} + \cdots + a_{rr}.$$

Note that

$$tr(\mathbf{A} + \mathbf{B}) = tr(\mathbf{A}) + tr(\mathbf{B}) \text{ and } tr(\mathbf{AB}) = tr(\mathbf{BA}).$$

The *matrix norm* $|\mathbf{A}|$ of an $r \times s$-matrix $\mathbf{A}$ may be written using the trace as

$$|\mathbf{A}| = \left[\sum_{i=1}^{r} \sum_{j=1}^{s} a_{ij}^2 \right]^{1/2} = \left[tr(\mathbf{AA}') \right]^{1/2} = \left[tr(\mathbf{A}'\mathbf{A}) \right]^{1/2}.$$

Eigenvalue decompositions: Let $\mathbf{A}$ be an $r \times r$ symmetric matrix. Then one can write that

$$\mathbf{A} = \mathbf{UDU}',$$

where $\mathbf{U}$ is an orthogonal matrix and $\mathbf{D}$ is a diagonal matrix with ordered diagonal elements $d_1 \geq d_2 \geq \cdots \geq d_r$. The columns of $\mathbf{U} = (\mathbf{u}_1 \cdots \mathbf{u}_r)$ are called the eigenvectors and the diagonal elements $d_1, ..., d_r$ corresponding eigenvalues of $\mathbf{A}$. Note that then it is true that $\mathbf{Au}_i = d_i \mathbf{u}_i$, $i = 1, ..., r$. Also note that $|\mathbf{A}|^2 = \sum d_i^2$.

Next let $\mathbf{A}$ be any $r \times s$ matrix, and assume that $r \leq s$. Then it can be written as

$$\mathbf{A} = \mathbf{UDV}',$$

where $\mathbf{U}$ and $\mathbf{V}$ are $r \times r$ and $s \times s$ orthogonal matrices, respectively, and $\mathbf{D} = (\mathbf{D}_1, \mathbf{D}_2)$ with $r \times r$ diagonal matrix $\mathbf{D}_1$ and $r \times s$ zero matrix $\mathbf{D}_1$. Again write $d_1 \geq \cdots \geq d_r$ for the diagonal elements of $\mathbf{D}_1$. The columns of $\mathbf{U}$ ($\mathbf{V}$) are the eigenvectors of $\mathbf{AA}'$ ($\mathbf{A}'\mathbf{A}$) with the first r eigenvalues d_i^2, $i = 1, ..., r$. Again, $|\mathbf{A}|^2 = \sum d_i^2$.

An $r \times r$ matrix $\mathbf{A}$ is said to be *nonnegative definite* if $\mathbf{a}'\mathbf{Aa} \geq 0$ for all r-vectors $\mathbf{a}$. We then write $\mathbf{A} \geq 0$. Moreover, $\mathbf{A}$ is *positive definite* (write $\mathbf{A} > 0$) if $\mathbf{a}'\mathbf{Aa} > 0$ for all r-vectors $\mathbf{a} \neq \mathbf{0}$.

The *Kronecker product* of two matrices $\mathbf{A}$ and $\mathbf{B}$, written $\mathbf{A} \otimes \mathbf{B}$, is the partitioned matrix

$$\mathbf{A} \otimes \mathbf{B} = \begin{pmatrix} a_{11}\mathbf{B} \; a_{21}\mathbf{B} \; \ldots \; a_{r1}\mathbf{B} \\ a_{12}\mathbf{B} \; a_{22}\mathbf{B} \; \ldots \; a_{r2}\mathbf{B} \\ \ldots \quad \ldots \quad \ldots \\ a_{1s}\mathbf{B} \; a_{2s}\mathbf{B} \; \ldots \; a_{rs}\mathbf{B} \end{pmatrix}.$$

The Kronecker product satisfies

$$(\mathbf{A} \otimes \mathbf{B})' = \mathbf{A}' \otimes \mathbf{B}' \quad \text{and} \quad (\mathbf{A} \otimes \mathbf{B})^{-1} = \mathbf{A}^{-1} \otimes \mathbf{B}^{-1}$$

and

$$(\mathbf{A} \otimes \mathbf{B})(\mathbf{C} \otimes \mathbf{D}) = (\mathbf{AC}) \otimes (\mathbf{BD}).$$

In statistics, people often wish to work with vectors instead of matrices. The *"vec" operation* is then used to vectorize a matrix. If $\mathbf{A} = (\mathbf{a}_1 \cdots \mathbf{a}_s)$ is an $r \times s$ matrix, then

$$\text{vec}(\mathbf{A}) = \begin{pmatrix} \mathbf{a}_1 \\ \ldots \\ \mathbf{a}_s \end{pmatrix}$$

just stacks the columns of $\mathbf{A}$ on top of each other. An often very useful result for vectorizing the product of three matrices is

$$\text{vec}(\mathbf{BCD}) = (\mathbf{D}' \otimes \mathbf{B}) \, \text{vec}(\mathbf{C}).$$

Clearly then, with two matrices $\mathbf{B}$ and $\mathbf{C}$,

$$\text{vec}(\mathbf{BC}) = (\mathbf{I} \otimes \mathbf{B}) \, \text{vec}(\mathbf{C}) = (\mathbf{C}' \otimes \mathbf{I}) \, \text{vec}(\mathbf{B}).$$

Note also that, for a $r \times r$ matrix $\mathbf{A}$, $tr(\mathbf{A}) = [\text{vec}(\mathbf{I}_r)]'\text{vec}(\mathbf{A})$.

Starting with vectorized $p \times p$ matrices, the following matrices prove very useful. Let $\mathbf{e}_i$ be a p-vector with ith element one and others zero. Then define the following $p^2 \times p^2$ matrices.

$$\mathbf{D}_{p,p} = \sum_{i=1}^{n} (\mathbf{e}_i \mathbf{e}_i') \otimes (\mathbf{e}_i \mathbf{e}_i'),$$

$$\mathbf{J}_{p,p} = \sum_{i=1}^{p} \sum_{j=1}^{p} (\mathbf{e}_i \mathbf{e}_j') \otimes (\mathbf{e}_i \mathbf{e}_j') = \text{vec}(\mathbf{I}_p)[\text{vec}(\mathbf{I}_p)]',$$

and

$$\mathbf{K}_{p,p} = \sum_{i=1}^{p} \sum_{j=1}^{p} (\mathbf{e}_i \mathbf{e}_j') \otimes (\mathbf{e}_j \mathbf{e}_i').$$

Note also that

$$\mathbf{I}_{p^2} = \sum_{i=1}^{p} \sum_{j=1}^{p} (\mathbf{e}_i \mathbf{e}_i') \otimes (\mathbf{e}_j \mathbf{e}_j').$$

Then one can easily see that, if $\mathbf{A}$ is a $p \times p$-matrix,

$$\mathbf{D}_{p,p}\text{vec}(\mathbf{A}) = \text{vec}(\text{diag}(\mathbf{A})) \quad \text{and} \quad \mathbf{J}_{p,p}\text{vec}(\mathbf{A}) = tr(\mathbf{A})\text{vec}(\mathbf{I}_p)$$

and

$$\mathbf{K}_{p,p}\text{vec}(\mathbf{A}) = \text{vec}(\mathbf{A}').$$

The matrix $\mathbf{K}_{p,p}$ is sometimes called a *commutation matrix*.

Appendix B
Asymptotical results for methods based on spatial signs

B.1 Some auxiliary results

Let $\mathbf{y} \neq \mathbf{0}$ and μ be any p-vectors, $p > 1$. Write also

$$r = |\mathbf{y}| \quad \text{and} \quad \mathbf{u} = |\mathbf{y}|^{-1}\mathbf{y}.$$

Then accuracies of different (constant, linear, and quadratic) approximations of function $|\mathbf{y} - \mu|$ of μ are

1.
$$||\mathbf{y} - \mu| - |\mathbf{y}|| \leq |\mu|.$$

2.
$$||\mathbf{y} - \mu| - |\mathbf{y}| - \mathbf{u}'\mu| \leq 2\frac{|\mu|^2}{r}.$$

See (3.8) in Bai et al. (1990).

3.
$$\left| |\mathbf{y} - \mu| - |\mathbf{y}| - \mathbf{u}'\mu - \mu'\frac{1}{2r}[\mathbf{I}_p - \mathbf{u}\mathbf{u}']\mu \right| \leq C\frac{|\mu|^{2+\delta}}{r^{1+\delta}}$$

for all $0 < \delta < 1$ where C does not depend on $\mathbf{y}$ or μ. Use Part 2 above and the Taylor theorem, and the result that

$$\min\{a, a^2\} \leq a^{1+\delta}, \qquad \text{for all } a > 0 \text{ and } 0 < \delta < 1.$$

See also Lemma 19(iv) in Arcones (1998).

In a similar way, the accuracies of constant and linear approximations of function $|\mathbf{y} + \mu|^{-1}(\mathbf{y} + \mu)$ of μ are given by

1.
$$\left| \frac{\mathbf{y} - \mu}{|\mathbf{y} + \mu|} - \frac{\mathbf{y}}{|\mathbf{y}|} \right| \leq 2\frac{|\mu|}{r}.$$

See (2.16) in Bai et al. (1990).

2.
$$\left| \frac{\mathbf{y}-\mu}{|\mathbf{y}+\mu|} - \frac{\mathbf{y}}{|\mathbf{y}|} - \frac{1}{r}[\mathbf{I}_p - \mathbf{u}\mathbf{u}']\mu \right| \leq C\frac{|\mu|^{1+\delta}}{r^{1+\delta}}$$

for all $0 < \delta < 1$ where C does not depend on $\mathbf{y}$ or μ. Use Part 1 above and the Taylor theorem again.

B.2 Basic limit theorems

For the first three theorems, see Sections 1.8 and 1.9 in Serfling (1980), for example.

Theorem B.1. *(Chebyshev). Let $y_1, y_2, \ldots$ be uncorrelated (univariate) random variables with means $\mu_1, \mu_2, \ldots$ and variances $\sigma_1^2, \sigma_2^2, \ldots$. If $\sum_{i=1}^n \sigma_i^2 = o(n^2)$, $n \to \infty$, then*

$$\frac{1}{n}\sum_{i=1}^n y_i - \frac{1}{n}\sum_{i=1}^n \mu_i \to_P 0.$$

Theorem B.2. *(Kolmogorov). Let $y_1, y_2, \ldots$ be independent (univariate) random variables with means $\mu_1, \mu_2, \ldots$ and variances $\sigma_1^2, \sigma_2^2, \ldots$. If $\sum_{i=1}^n \sigma_i^2/i^2$ converges, then*

$$\frac{1}{n}\sum_{i=1}^n y_i - \frac{1}{n}\sum_{i=1}^n \mu_i \to 0 \text{ almost surely.}$$

The following theorem (Corollary A.1.2 in Hettmansperger and McKean (1998) will be useful in the future. Note that the simple central limit theorem is a special case.

Theorem B.3. *Suppose that (univariate) random variables $y_1, y_2, \ldots$ are iid with $E(y_i) = 0$ and $Var(y_i) = 1$. Suppose that the triangular array of constants $c_{1n}, \ldots, c_{nn}$, $n = 1, 2, \ldots$, is such that*

$$\sum_{i=1}^n c_{in}^2 \to \sigma^2, \quad 0 < \sigma^2 < \infty$$

and

$$\max_{1 \leq i \leq n} |c_{in}| \to 0, \quad \text{as } n \to \infty.$$

Then

$$\sum_{i=1}^n c_{in} y_i \to_D N(0, \sigma^2).$$

The following result is an easy consequence of Corollary 1.9.2 in Serfling (1980).

Theorem B.4. *Suppose that $y_1, y_2, \ldots$ are independent random variables with $E(y_i) = 0$, $E(y_i^2) = \sigma_i^2$ and $E(|y_i|^3) = \gamma_i < \infty$. If*

$$\frac{(\sum_{i=1}^n \gamma_i)^2}{(\sum_{i=1}^n \sigma_i^2)^3} \to 0$$

then

$$\frac{\sum_{i=1}^n y_i}{\sqrt{\sum_{i=1}^n \sigma_i^2}} \to_d N(0,1).$$

The following key result is Lemma 4.2 in Davis et al. (1992) and Theorem 1 in Arcones (1998).

Theorem B.5. *Let $G_n(\mu)$, $\mu \in \Re^p$, be a sequence of convex stochastic processes, and let $G(\mu)$ be a convex (limit) process in the sense that the finite dimensional distributions of $G_n(\mu)$ converge to those of $G(\mu)$. Let $\hat{\mu}, \hat{\mu}_1, \hat{\mu}_2, \ldots$ be random variables such that*

$$G(\hat{\mu}) = \inf_\mu G(\mu) \quad \text{and} \quad G_n(\hat{\mu}_n) = \inf_\mu G_n(\mu), \quad n = 1, 2, \cdots.$$

Then

$$\hat{\mu}_n \to_d \hat{\mu}.$$

B.3 Notation and assumptions

Let $\mathbf{y}$ be a p-variate random vector with cdf F and $p > 1$. The *spatial median* of F minimizes the objective function

$$D(\mu) = E\{|\mathbf{y} - \mu| - |\mathbf{y}|\}.$$

(Note that no moment assumptions are needed as $D(\mu) \leq |\mu|$.) We wish to test the null hypothesis $H_0 : \ \mu = 0$ and also estimate the unknown value of μ.

Let $\mathbf{y}_1, \ldots, \mathbf{y}_n$ be a random sample from a p-variate distribution F, $p > 1$. Write

$$D_n(\mu) = \text{ave}\{|\mathbf{y}_i - \mu| - |\mathbf{y}_i|\}.$$

The function $D_n(\mu)$ as well as $D(\mu)$ is convex and bounded. The *sample spatial median* is defined as

$$\hat{\mu} = \arg\min D_n(\mu).$$

We also define vector- and matrix-valued functions

$$\mathbf{U}(\mathbf{y}) = \frac{\mathbf{y}}{|\mathbf{y}|}, \quad \mathbf{A}(\mathbf{y}) = \frac{1}{|\mathbf{y}|}\left[\mathbf{I}_p - \frac{\mathbf{y}\mathbf{y}'}{|\mathbf{y}|^2}\right], \quad \text{and} \quad \mathbf{B}(\mathbf{y}) = \frac{\mathbf{y}\mathbf{y}'}{|\mathbf{y}|^2}$$

for $\mathbf{y} \neq \mathbf{0}$ and $\mathbf{U}(\mathbf{0}) = \mathbf{0}$ and $\mathbf{A}(\mathbf{0}) = \mathbf{B}(\mathbf{0}) = \mathbf{0}$. The statistic

$$\mathbf{T}_n = \text{ave}\{\mathbf{U}(\mathbf{y}_i)\}$$

is then the *spatial sign test statistic* for testing the null hypothesis that the spatial median is zero.

Assumption 11 *We assume that (i) the density function f of $\mathbf{y}$ is continuous and bounded in an open neighborhood of the origin, and that (ii) the spatial median of the distribution of $\mathbf{y}$ is zero and unique; that is,*

$$D(\mu) > 0, \quad \forall \mu \neq \mathbf{0}.$$

First note the following.

Lemma B.1. *If the density function f of $\mathbf{y}$ is continuous and bounded in an open neighborhood of the origin then $E(|\mathbf{y}|^{-\alpha}) < \infty$ for all $0 \leq \alpha < 2$.*

We also write

$$\mathbf{A} = E\{\mathbf{A}(\mathbf{y})\} \quad \text{and} \quad \mathbf{B} = E\{\mathbf{B}(\mathbf{y})\}.$$

The expectation defining $\mathbf{B}$ clearly exists and is bounded ($|\mathbf{B}(\mathbf{y})| = 1$). Our assumption implies that $E(|\mathbf{y}|^{-1}) < \infty$ and therefore also $\mathbf{A}$ exists and is bounded. Auxiliary results in Section B.1 and Lemma B.1 then imply the following.

Lemma B.2. *Under our assumptions,*

$$D(\mu) = \frac{1}{2}\mu'\mathbf{A}\mu + o(|\mu|^2).$$

See also Lemma 19 in Arcones (1998).

B.4 Limiting results for spatial median

Lemma B.3. $\hat{\mu} \to \mathbf{0}$ *almost surely.*

Proof.

1. First, $D_n(\mu) \to D(\mu)$ almost surely for all μ (LLN).
2. As $D_n(\mu)$ and $D(\mu)$ are bounded and convex, also

$$\sup_{|\mu| \leq C} |D_n(\mu) - D(\mu)| \to 0 \text{ almost surely}$$

for all $C > 0$ (Theorem 10.8 in Rockafellar (1970)).

3. Write

$$\hat{\mu}^* = \arg \min_{|\mu| \leq C} D_n(\mu)$$

for some C. Then $D_n(\hat{\mu}^*) \to 0$ and $\hat{\mu}^* \to \mathbf{0}$ almost surely. This is seen as

$$D_n(\hat{\mu}^*) \leq D_n(\mathbf{0}) \to D(\mathbf{0}) \leq D(\hat{\mu}^*)$$

almost surely and $D_n(\hat{\mu}^*) - D(\hat{\mu}^*) \to 0$ almost surely.

4. Finally, we show that $|\hat{\mu}| \le C$ almost surely. Write

$$\delta = \inf_{|\mu| \ge C} D(\mu).$$

If $\mu \ge C$ then, for any $0 < \varepsilon < \delta$,

$$D_n(\mu) > \delta - \varepsilon > D_n(\mathbf{0}) \ge D_n(\hat{\mu}^*)$$

almost surely, and the result follows.

Using our results in Sections B.1 and B.2 we easily get the following.

Lemma B.4. *Under our assumptions,* $\sqrt{n}\mathbf{T}_n \to_d N_p(\mathbf{0}, \mathbf{B})$.

and

Lemma B.5. *Under our assumptions*

$$nD_n(n^{-1/2}\mu) - \left(\sqrt{n}\mathbf{T}_n - \frac{1}{2}\mathbf{A}\mu\right)' \mu \to_P 0.$$

Then Theorem B.5 implies the next theorem.

Theorem B.6. *Under our assumptions,*

$$\sqrt{n}\hat{\mu} \to_d N_p(\mathbf{0}, \mathbf{A}^{-1}\mathbf{B}\mathbf{A}^{-1}).$$

The proof was constructed in the multivariate case ($p > 1$). The univariate case can be proved in the same way. The matrix $\mathbf{A}$ is then replaced by the scalar $a = 2f(0)$ and $\mathbf{B}$ by $b = 1$.

B.5 Limiting results for the multivariate regression estimate

We now assume that

$$\mathbf{y}_i = \beta' \mathbf{x}_i + \varepsilon_i, \quad i = 1, ..., n,$$

where $\varepsilon_1, ..., \varepsilon_n$ are iid from a distribution satisfying Assumption 11. The q-variate design variables satisfy the following assumption.

Assumption 12 *Let* $\mathbf{X} = (\mathbf{x}_1, ..., \mathbf{x}_n)'$ *be the (fixed)* $n \times q$ *design matrix. We assume that*

$$\frac{1}{n}\mathbf{X}'\mathbf{X} \to \mathbf{D},$$

and

$$\frac{\max_{1 \le i \le n}\{\mathbf{x}_i'\mathbf{C}'\mathbf{C}\mathbf{x}_i\}}{\sum_{i=1}^n \{\mathbf{x}_i'\mathbf{C}'\mathbf{C}\mathbf{x}_i\}} \to 0 \quad \textit{for all } p \times q \textit{ matrices } \mathbf{C}.$$

It is not a restriction to assume that $\beta = \mathbf{0}$ so that $\mathbf{y}_1,...,\mathbf{y}_n$ also satisfy Assumption 11. Write

$$\mathbf{T}_n = \text{ave}\left\{\mathbf{u}_i\mathbf{x}_i'\right\}$$

which can be used to test the null hypothesis $H_0 : \beta = \mathbf{0}$. Consider now the objective function

$$D_n(\beta) = \mathbf{AVE}\{|\mathbf{y}_i - \mu_i| - |\mathbf{y}_i|\},$$

where

$$\mu_i = \beta'\mathbf{x}_i, \quad i = 1,...,n.$$

The function $D_n(\beta)$ is again convex, and the solution is unique if there is no $p \times q$-matrix $\mathbf{C}$ such that $\mathbf{y}_i = \mathbf{C}\mathbf{x}_i$ for all $i = 1,...,n$. The *multivariate regression estimate* is then defined as

$$\hat{\beta} = \text{arg min } D_n(\beta).$$

Again, using our results in Sections B.1 and B.2, we obtain the following.

Lemma B.6. *Under our assumptions,* $\sqrt{n}\, vec(\mathbf{T}_n) \to_d N_p(\mathbf{0}, \mathbf{D} \otimes \mathbf{B})$.

Lemma B.7. *Under our assumptions,*

$$nD_n(n^{-1/2}\beta) - vec\left(\sqrt{n}\mathbf{T}_n - \frac{1}{2}\mathbf{A}\beta\mathbf{D}\right)' vec(\beta) \to_P 0.$$

Then Theorem B.5 implies the following.

Theorem B.7. *Under our assumptions,*

$$\sqrt{n} vec(\hat{\beta}) \to_d N_{pq}(\mathbf{0}, \mathbf{D}^{-1} \otimes \mathbf{A}^{-1}\mathbf{B}\mathbf{A}^{-1}).$$

References

Anderson, T.W. (1999). Asymptotic theory for canonical correlation analysis. *Journal of Multivariate Analysis*, **70**, 1–29.

Anderson, T.W. (2003). *An Introduction to Multivariate Statistical Analysis*. Third Edition, Wiley, New York.

Arcones, M.A. (1998). Asymptotic theory for M-estimators over a convex kernel. *Econometric Theory,* **14**, 387–422.

Arcones, M.A., Chen, Z., and Gine, E. (1994). Estimators related to U-processes with applications to multivariate medians: Asymptotic normality. *Annals of Statistics*, **22**, 1460–1477.

Azzalini, A. (2005). The skew-normal distribution and related multivariate families. *Scandinavian Journal of Statistics*, **32**, 159–188.

Bai, Z.D., Chen, R., Miao, B.Q., and Rao, C.R. (1990). Asymptotic theory of least distances estimate in multivariate linear models. *Statistics*, **4**, 503–519.

Barnett, V. (1976). The ordering of multivariate data. *Journal of Royal Statistical Society, A,* **139**, 318–355.

Bassett, G. and Koenker, R. (1978). Asymptotic theory of least absolute error regression. *Journal of the American Statistical Association*, **73**, 618–622.

Bickel, P.J. (1964). On some asymptotically nonparametric competitors of Hotelling's T^2. *Annals of Mathematical Statistics*, **36**, 160–173.

Bilodeau, M. and Brenner, D. (1999). *Theory of Multivariate Statistics.* Springer-Verlag, New York.

Blomqvist, N. (1950). On a measure of dependence between two random variables. *Annals of Mathematical Statistics*, **21**, 593–600.

Blumen, I. (1958). A new bivariate sign test for location. *Journal of the American Statistical Association*, **53**, 448–456.

Brown (1983). Statistical uses of the spatial median, *Journal of the Royal Statistical Society, B*, **45**, 25–30.

Brown, B. and Hettmansperger, T. (1987). Affine invariant rank methods in the bivariate location model. *Journal of the Royal Statististical Society, B* **49**, 301–310.

Brown, B. and Hettmansperger, T. (1989). An affine invariant bivariate version of the sign test. *Journal of the Royal Statistical Society, B* **51**, 117–125.

Brown, B.M., Hettmansperger, T.P., Nyblom, J., and Oja, H. (1992). On certain bivariate sign tests and medians. *Journal of the American Statistical Association*, **87**, 127–135.

Busarova, D., Tyurin, Y., Möttönen, J., and Oja, H. (2006). Multivariate Theil estimator with the corresponding test. *Mathematical Methods of Statistics*, **15**, 1–19.

Chakraborty, B. (1999). On multivariate median regression. *Bernoulli*, **5**, 683–703.

Chakraborty, B. (2003). On multivariate quantile regression. *Journal of Statistical Planning and Inference*, **110**, 109–132.

Chakraborty, B. and Chaudhuri, P. (1996). On the transformation and retransformation technique for constructing affine equivariant multivariate median. *Proceedings of the American Mathematical Society*, **124**, 2359–2547.

Chakraborty, B. and Chaudhuri, P. (1998). On an adaptive transformation retransformation estimate of multivariate location. *Journal of the Royal Statistical Society, B.*, **60**, 147–157.

Chakraborty, B. and Chaudhuri, P. (1999). On affine invariant sign and rank tests in one sample and two sample multivariate problems. In: *Multivariate, Design and Sample Survey* (ed. S. Ghosh). Marcel-Dekker, New York. pp. 499–521.

Chakraborty, B., Chaudhuri, P., and Oja, H. (1998). Operating transformation retransformation on spatial median and angle test. *Statistica Sinica*, **8**, 767–784.

Chaudhuri, P. (1992). Multivariate location estimation using extension of R-estimates through U-statistics type approach, *Annals of Statistics*, **20**, 897–916.

Chaudhuri, P. (1996). On a geometric notion of quantiles for multivariate data. *Journal of the American Statistical Society*, **91**, 862–872.

Chaudhuri, P. and Sengupta, D. (1993). Sign tests in multidimension: Inference based on the geometry of data cloud. *Journal of the American Statistical Society*, **88**, 1363–1370.

Choi, K. and Marden, J. (1997). An approach to multivariate rank tests in multivariate analysis of variance. *Journal of the American Statistical Society*, **92**, 1581–1590.

Croux, C. and Haesbrock, G. (2000). Principal component analysis based on robust estimators of the covariance or correlation matrix: Influence functions and efficiencies. *Biometrika*, **87**, 603–618.

Croux, C., Ollila, E. and Oja, H. (2002). Sign and rank covariance matrices: Statistical properties and application to principal component analysis. In *Statistical Data Analysis Based on the L1-Norm and Related Methods* (ed. by Yadolah Dodge) Birkhäuser, Basel, pp. 257–270.

Dalgaard, P. (2008). *Introductory Statistics with R*, Second Edition, Springer, New York.

DasGupta, S. (1999a). Lawley-Hotelling trace. In: *Encyclopedia of Biostatistics*, Wiley, New York.

DasGupta, S. (1999b). Wilks' Lambda criterion. In: *Encyclopedia of Biostatistics*, Wiley, New York.

Datta, S. and Satten, G. A. (2005). Rank-sum tests for clustered data. *Journal of the American Statistical Association*, **100**, 908–915.

Datta, S. and Satten, G. A. (2008). A signed-rank test for clustered data. *Biometrics*, **64**, 501–507.

Davies, P.L. (1987). Asymptotic behavior of S-estimates of multivariate location parameters and dispersion matrices. *Annals of Statistics*, **15**, 1269–1292.

Davis, J. B. and McKean, J. (1993). Rank-based methods for multivariate linear models. *Journal of the American Statistical Association*, **88**, 245–251.

Davis, R. A., Knight, K. and Liu, J. (1992). M-estimation for autoregression with infinite variance. *Stochastic Processes and Their Applications*, **40**, 145–180.

Dietz, E.J. (1982). Bivariate nonparametric tests for the one-sample location problem. *Journal of the American Statistical Association* **77**, 163–169.

Donoho, D.L. and Huber, P.J. (1983). The notion of breakdown poin. In: *A Festschrift for Erich L. Lehmann* (ed. P.J. Bickel, K.A. Doksum and J.L. Hodges) Belmont, Wadsworth, pp. 157–184.

Dümbgen, L. (1998). On Tyler's M-functional of scatter in high dimension. *Annals of the Institute of Statistal Mathematics*, **50**, 471–491.

Dümbgen, L. and Tyler, D. (2005). On the breakdown properties of some multivariate M-functionals. *Scandinavian Journal of Statistics*, **32**, 247–264.

Everitt, B. (2004). *An R and S-PLUS Companion to Multivariate Analysis*. London: Springer.

Frahm, G. (2004). *Generalized Elliptical Distributions: Theory and Applications.* Doctoral Thesis: Universität zu Köln, Wirtschafts- uund Sozialwissenschaftliche Fakultät. Seminar für Wirtschafts- und Sozialstatistik.

Gieser, P.W. and Randles, R.H. (1997). A nonparametric test of independence between two vectors. *Journal of the American Statistical Association*, **92**, 561–567.

Gini and Galvani (1929). Di talune estensioni dei concetti di media ai caratteri qualitative. *Metron*, **8**

Gómez, E., Gómez-Villegas, M.A., and Marín, J.M. (1998). A multivariate generalization of the power exponential family of distributions. *Communications in Statististics -Theory and Methods*, **27**, 3, 589–600.

Gower, J. S. (1974). The mediancentre. *Applied Statistics*, **2**, 466–470.

Haataja, R., Larocque, D., Nevalainen, J., and Oja, H. (2008). A weighted multivariate signed-rank test for cluster-correlated data. *Journal of Multivariate Analysis*, **100**, 1107–1119.

Haldane, J.B.S. (1948). Note on the median of the multivariate distributions. *Biometrika*,**35**, 414–415.

Hallin, M. and Paindaveine, D. (2002). Optimal tests for multivariate location based on interdirections and pseudo-Mahalanobis ranks. *Annals of Statistics*, **30**, 1103–1133.

Hallin, M. and Paindaveine, D. (2006). Semiparametrically efficient rank-based inference for shape. I. Optimal rank-based tests for sphericity. *Annals of Statistics*, **34**, 2707–2756.

Hampel, F.R. (1968). *Contributions to the theory of robust estimation*. Ph.D. Thesis, University of California, Berkeley.

Hallin, M., Oja, H., and Paindaveine, D. (2006). Semiparametrically efficient rank-based inference for shape. II. Optimal R-estimation of shape. *Annals of Statistics*, **34**, 2757–2789.

Hampel, F. R. (1974). The influence curve and its role in robust estimation. *Journal of the American Statistical Association*, **62**, 1179–1186.

Hampel, F.R., Rousseeuw, P.J., Ronchetti, E.M., and Stahel, W.A. (1986). *Robust Statistics: The Approach Based on Influence Functions*. Wiley, New York.

Hettmansperger, T.P. and Aubuchon, J.C. (1988). Comment on "Rank-based robust analysis of linear models. I. Exposition and review" by David Draper. *Statistical Science*, **3**, 262–263.

Hettmansperger, T.P. and McKean, J.W. (1998). *Robust Nonparametric Statistical Methods*. Arnold, London.

Hettmansperger, T.P. and Oja, H. (1994). Affine invariant multivariate multisample sign tests. *Journal of the Royal Statistical Society, B*, **56**, 235–249.

Hettmansperger, T.P. and Randles, R.H. (2002). A practical affine equivariant multivariate median. *Biometrica*, **89**, 851–860.

Hettmansperger, T. P. , Möttönen, J., and Oja, H. (1997). Multivariate affine invariant one-sample signed-rank tests. *Journal of the American Statistical Association*, **92**, 1591–1600.

Hettmansperger, T. P. , Möttönen, J., and Oja, H. (1998). Multivariate affine invariant rank tests for several samples. *Statistica Sinica*, **8**, 785–800.

Hettmansperger, T. P. , Möttönen, J., and Oja, H. (1999). The geometry of the affine invariant multivariate sign and rank methods. *Journal of Nonparametric Statistics*, **11**, 271–285.

Hettmansperger, T.P., Nyblom, J., and Oja, H. (1992). On multivariate notions of sign and rank. *L1-Statistical Analysis and Related Methods,* 267–278. Ed. by Y. Dodge. Elsevier, Amsterdam.

Hettmansperger, T.P., Nyblom , J., and Oja, H. (1994). Affine invariant multivariate one-sample sign test. *Journal of the Royal Statistical Society, B*, **56**, 221–234.

Hodges, J.L. (1955). A bivariate sign test. *Ann. Math. Statist.*, **26**, 523–527.

Hollander, M. and Wolfe, D.A. (1999). *Nonparametric Statistical Methods*. 2nd ed. Wiley, New York.

Huber, P.J. (1981). *Robust Statistics*. Wiley, New York.

Hyvärinen, A., Karhunen, J., and Oja, E. (2001). *Independent Component Analysis*. Wiley, New York.

Hössjer, O. and Croux, C. (1995). Generalizing univariate signed rank statistics for testing and estimating a multivariate location parameter. *Journal of Nonparametric Statistics*, **4**, 293–308.

Jan, S.L. and Randles, R.H. (1994). A multivariate signed sum test for the one sample location problem. *Journal of Nonparametric Statistics*, **4**, 49–63.

Jan, S.L. and Randles, R.H. (1996). Interaction tests for simple repeated measures designs. *Journal of the American Statistical Association*, **91**, 1611–1618.

John, S. (1971). Some optimal multivariate tests. *Biometrika*, **58**, 123–127.

John, S. (1972). The distribution of a statistic used for testing sphericity of normal distributions. *Biometrika*, **59**, 169–173.

Johnson, R.A. and Wichern, D.W. (1998). *Multivariate Statistical Analysis*. Prentice Hall, Upper Saddle River, NJ.

Kankainen, A., Taskinen, S., and Oja, H. (2007). Tests of multinormality based on location vectors and scatter matrices. *Statistical Methods & Applications*, **16**, 357–379.

Kendall, M.G. (1938). A new measure of rank correlation. *Biometrika*, **30**, 81–93.

Kent, J. and Tyler, D. (1996). Constrained M-estimation for multivariate location and scatter. *Annals of Statistics*, **24**, 1346–1370.

Koltchinskii, V. I. (1997). M-estimation, convexity and quantiles. *Annals of Statistics*, **25**, 435–477.

Koshevoy, G. and Mosler, K. (1997a). Multivariate Gini indices. *Journal of Multivariate Analysis*, **60**, 252–276.

Koshevoy, G. and Mosler, K. (1997b). Zonoid trimming for multivariate distributions. *Annals of Statistics*, **25**, 1998–2017.

Koshevoy, G. and Mosler, K. (1998). Lift zonoids, random convex hulls and the variability of random vectors. *Bernoulli*, **4**, 377–399.

Koshevoy, G., Möttönen, J., and Oja, H. (2003). A scatter matrix estimate based on the zonotope. *Annals of Statistics*, **31**, 1439–1459.

Koshevoy, G., Möttönen, J., and Oja, H. (2004). On the geometry of multivariate L_1 objective functions. *Allgemeines Statistisches Archiv*, **88**, 137–154.

Larocque, D. (2003). An affine-invariant multivariate sign test for cluster correlated data. *The Canadian Journal of Stastistics*, **31**, 437–455.

Larocque, D., Nevalainen, J., and Oja, H. (2007). A weighted multivariate sign test for cluster correlated data. *Biometrika*, **94**, 267–283.

Larocque, D., Nevalainen, J. and Oja, H. (2007). One-sample location tests for multilevel data. *Journal of Statistical Planning and Inference*, **8**, 2469–2482.

Larocque, D., Haataja, R., Nevalainen, J., and Oja, H. (2010). Two sample tests for the nonparametric Behrens-Fisher problem with clustered data. *Journal of Nonparametric Statistics*, to appear.

Larocque, D., Tardif, S. and van Eeden, C. (2000). Bivariate sign tests based on the sup, L1 and L2-norms. *Annals of the Institute of Statistical Mathematics* **52**, 488–506.

Lehmann, E.L. (1998). *Nonparametrics: Statistical Methods Based on Ranks*. Prentice Hall, Englewood Cliffs, NJ.

Liu, R. Y.(1990). On a notion of data depth based upon random simplices. *Annals of Statistics*, **18**, 405–414.

Liu, R. Y. (1992). Data depth and multivariate rank tests. In: *L1-Statistical Analysis and Related Methods* (ed. Y. Dodge), Elsevier, Amsterdam, pp. 279–294..

Liu, R. and Singh, K. (1993). A quality index based on data depth and multivariate rank tests. *Journal of the American Statistical Association*, **88**, 252–260.

Liu, R. Parelius, J.M., and Singh, K. (1999). Multivariate analysis by data depth: Descriptive statistics, graphics and inference (with discussion). *Annals of Statistics*, **27**, 783–858.

Locantore, N., Marron, J.S., Simpson, D.G., Tripoli, N., Zhang, J.T., and Cohen, K.L. (1999). Robust principal components for functional data. *Test*, **8**, 1–73.

Lopuhaä, H.P. (1989). On the relation between S-estimators and M-estimators of multivariate location and covariance. *Annals of Statistics*, **17**, 1662–1683.

Lopuhaä, H.P. and Rousseeuw, P.J. (1991). Breakdown properties of affine equivariant estimators of multivariate location and covariance matrices. *Annals of Statistics*, **19**, 229–248.

Magnus, J.R. and Neudecker, H. (1988). *Matrix Differential Calculus with Applications in Statistics and Economics*. Wiley, New York.

Mardia, K.V. (1967). A nonparametric test for the bivariate location problem. *Journal of the Royal Statistical Society, B*, **290**, 320–342.

Mardia, K.V. (1972). *Statistics of Directional Data*. Academic Press, London.

Mardia, K.V., Kent, J.T., and Bibby, J.M. (1979). *Multivariate Analysis*. Academic Press, Orlando, FL.

Marden, J. (1999a). Multivariate rank tests. In *Design of Experiments, and Survey Sampling* (ed. S. Ghosh), M. Dekker, New York, pp. 401–432.

Marden, J. (1999b). Some robust estimates of principal components. *Statistics & Probability Letters*, **43**, 349–359.

Maronna, R.A. (1976). Robust M-estimators of multivariate location and scatter. *Annals of Statistics*, **4**, 51–67.

Maronna, R.A., Martin, R.D., and Yohai, V. Y. (2006). *Robust Statistics. Theory and Methods*. Wiley, New York.

Mauchly, J.W. (1940). Significance test for sphericity of a normal *n*-variate distribution. *Annals of Mathematical Statistics*, **11**, 204–209.

McLachlan, G. and Peel, D. (2000). *Finite Mixture Models.* Wiley, New York.

Milasevic, P. and Ducharme, G.R. (1987). Uniqueness of the spatial median. *Annals of Statistics*, **15**, 1332–1333.

Mood, A.M. (1954). On the asymptotic efficiency of certain nonparametric two-sample tests. *Annals of Mathematical Statistics*, **25**, 514–533.

Mosler, K. (2002). *Multivariate Dispersion, Central Regions, and Depth: The Lift Zonoid Approach.* Springer, New York.

Möttönen, J. and Oja, H. (1995). Multivariate spatial sign and rank methods. *Journal of Nonparametric Statistics*, **5**, 201–213.

Möttönen, J., Hüsler, J., and Oja, H. (2003). Multivariate nonparametric tests in randomized complete block design. *Journal of Multivariate Analysis*, **85**, 106–129.

Möttönen, J., Hettmansperger, T.P., Oja, H., and Tienari, J. (1998). On the efficiency of the affine invariant multivariate rank tests. *Journal of Multivariate Analysis*, **66**, 118–132.

Möttönen, J., Oja, H., and Serfling, R. (2005). Multivariate generalized spatial signed-rank methods. *Journal of Statistical Research*, **39**, 19–42.

Möttönen, J., Oja, H., and Tienari, J. (1997). On the efficiency of multivariate spatial sign and rank tests. *Annals of Statistics*, **25**, 542–552.

Muirhead, R.J. (1982). *Aspects of Multivariate Statistical Theory*. Wiley, New York.

Muirhead, R.J. and Waternaux, C.M. (1980). Asymptotic distributions in canonical correlation analysis and other multivariate procedures for nonnormal populations. *Biometrika*, **67**, 31–43.

Nevalainen, J. and Oja, H. (2006). SAS/IML macros for a multivariate analysis of variance based on spatial signs. *Journal of Statistical Software*, **16**, Issue 5.

Nevalainen, J., Larocque, D., and Oja, H. (2007b). On the multivariate spatial median for clustered data. *Canadian Journal of Statistics*, **35**, 215–231.

Nevalainen, J, Larocque, D., and Oja, H. (2007a). A weighted spatial median for clustered data. *Statistical Methods & Applications*, **15**, 355–379.

Nevalainen, J., Larocque, D., Oja, H., and Pörsti, I. (2009). Nonparametric analysis of clustered multivariate data. *Journal of the American Statistical Association*, under revision.

Nevalainen, J., Möttönen, J., and Oja, H. (2007). A spatial rank test and corresponding rank estimates for several samples. *Statistics & Probability Letters*, **78**, 661–668.

Niinimaa, A. and Oja, H. (1995). On the influence function of certain bivariate medians. *Journal of the Royal Statistical Society, B*, **57**, 565–574.

Niinimaa, A. and Oja, H. (1999). Multivariate median. In: *Encyclopedia of Statistical Sciences (Update Volume 3)* (ed. S. Kotz, N.L. Johnson, and C.P. Read), Wiley, New York.

Nordhausen, K., Oja, H., and Paindaveine, D. (2009). Signed-rank tests for location in the symmetric independent component model. *Journal of Multivariate Analysis*, **100**, 821–834.

Nordhausen, K., Oja, H. and Ollila, E. (2009). Multivariate Models and the First Four Moments. In: *Festschrift in Honour of Tom Hettmansperger*, to appear.

Nordhausen, K., Oja, H., and Tyler, D. (2006). On the efficiency of invariant multivariate sign and rank tests. In: *Festschrift for Tarmo Pukkila on his 60th Birthday* (ed. E. Liski, J. Isotalo, J, J. Niemel, S. Puntanen, and G. Styan).

Oja, H. (1983). Descriptive statistics for multivariate distributions. *Statistics & Probability Letters* **1**, 327–332.

Oja, H. (1987). On permutation tests in multiple regression and analysis of covariance problems. *Australian Journal of Statistics* **29**, 81–100.

Oja, H. (1999). Affine invariant multivariate sign and rank tests and corresponding estimates: a review. *Scandinavian Journal of Statistics* **26**, 319–343.

Oja, H. and Niinimaa, A. (1985). Asymptotical properties of the generalized median in the case of multivariate normality. *Journal of the Royal Statistical Society, B*, **47**, 372–377.

Oja, H. and Nyblom, J. (1989). On bivariate sign tests. *Journal of the American Statistical Association*, **84**, 249–259.

Oja, H. and Paindaveine, D. (2005). Optimal signed-rank tests based on hyperplanes. *Journal of Statistical Planning and Inference*, **135**, 300–323.

Oja, H. and Randles, R.H. (2004). Multivariate nonparametric tests. *Statistical Science*, **19**, 598–605.

Oja, H., Paindaveine, D., and Taskinen, S. (2009). Parametric and nonparametric tests for multivariate independence in the independence component model. Submitted.

Oja, H., Sirkiä, S., and Eriksson, J. (2006). Scatter matrices and independent component analysis. *Austrian Journal of Statistics*, **35**, 175–189.

Ollila, E., Croux, C., and Oja, H. (2004). Influence function and asymptotic efficiency of the affine equivariant rank covariance matrix. *Statistica Sinica*, **14**, 297–316.

Ollila, E., Hettmansperger, T.P., and Oja, H. (2002). Estimates of regression coefficients based on sign covariance matrix. *Journal of the Royal Statistical Society, B*, **64**, 447–466.

Ollila, E., Oja, H., and Croux, C. (2003b). The affine equivariant sign covariance matrix: Asymptotic behavior and efficiency. *Journal of Multivariate Analysis*, **87**, 328–355.

Ollila, E., Oja, H., and Koivunen, V. (2003). Estimates of regression coefficients based on rank covariance matrix. *Journal of the American Statistical Association*, **98**, 90–98.

Paindaveine, D. (2008). A canonical definition of shape. *Statistics & Probability Letters*, **78**, 2240–2247.

Pesarin, F. (2001). *Multivariate Permutation Tests with applications in Biostatistics*, Wiley, Chichester.

Peters, D. and Randles, R.H. (1990). A multivariate signed-rank test for the one-sampled location problem. *Journal of the American Statistical Association*, **85**, 552–557.

Peters, D. and Randles, R.H. (1990). A bivariate signed rank test for the two-sample location problem. *Journal of the Royal Statistical Society, B*, **53**, 493–504.

Pillai, K.C.S. (1955). Some new test criteria in multivariate analysis. *Annals of Mathematical Statistics*, **26**, 117–121.

Portnoy, S. and Koenker, R. (1997). The Gaussian hare and the Laplacian tortoise: Computability of squared-error versus absolute-error estimators. *Statistical Science*, **12**, 279–300.

Puri, M.L. and Sen, P.K. (1971). *Nonparametric Methods in Multivariate Analysis*. Wiley, New York.

Puri, M.L. and Sen, P.K. (1985). *Nonparametric Methods in General Linear Models*. Wiley, New York.

Randles, R.H. (1989). A distribution-free multivariate sign test based on interdirections. *Journal of the American Statistical Association*, **84** 1045–1050.

Randles, R.H. (1992). A Two Sample Extension of the Multivariate Interdirection Sign Test. *L1-Statistical Analysis and Related Methods* (ed. Y. Dodge), Elsevier, Amsterdam, pp. 295–302.

Randles, R.H. (2000). A simpler, affine equivariant multivariate, distribution-free sign test. *Journal of the American Statistical Association*, **95**, 1263–1268.

Randles, R.H. and Peters, D. (1990). Multivariate rank tests for the two-sample location problem. *Communications in Statistics – Theory and Methods*, **19**, 4225–4238.

Rao, C.R. (1948). Tests of significance in multivariate analysis. *Biometrika* **35**, 58–79.

Rao, C.R. (1988). Methodology based on L_1-norm in statistcal inference. *Sankhya A* **50**, 289–311.

Rockafellar, R.T. (1970). *Convex Analysis*. Princeton University Press, Princeton, NJ.

Rosner, B. and Grove, D. (1999). Use of the Mann-Whitney U-test for clustered data. *Statistics in Medicine*, **18**, 1387–1400.

Rosner, B. Grove, D., and Ting Lee, M.-L. (2003). Incorporation of clustering effects for the Wilcoxon rank sum test: A large sample approach. *Biometrics*, **59**, 1089–1098.

Rosner, B. Grove, D., and Ting Lee, M.-L. (2006). The Wilcoxon signed rank test for paired comparisons of clustered data. *Biometrics*, **62**, 185–192.

Seber, G.A.F. (1984). *Multivariate Observations*. Wiley, New York.

Serfling, R.J. (1980). *Approximation Theorems of Mathematical Statistics*. Wiley, New York.

Serfling, R.J. (2004). Nonparametric multivariate descriptive measures based on spatial quantiles. *Journal of Statistical Planning and Inference*, **123**, 259–278.

Shen, G. (2008). Asymptotics of Oja median estimate. *Statistics and Probability Letters*, **78**, 2137–2141.

Shen, G. (2009). Asymptotics of a Theil-type estimate in multiple linear regression. *Statistics and Probability Letters*, **79**, 1053–1064.

Sirkiä, S., Taskinen, S., and Oja, H. (2007). Symmetrised M-estimators of scatter. *Journal of Multivariate Analysis*, **98**, 1611–1629.

Sirkiä, S., Taskinen, S., Oja, H. and Tyler, D. (2008). Tests and estimates of shape based on spatial signs and ranks. *Journal of Nonparametric Statistics*, **21**, 155–176.

Small, C.G. (1990). A survey of multidimensional medians. *International Statistical Review*, **58**, 263–277.

Spearman, C. (1904). The proof and measurement of association between two things. *American J. Psychology*, **15**, 72–101.

Spjøtvoll, E. (1968). A note on robust estimation in analysis of variance. *Annals of Mathematical Statistics*, **39**, 1486–1492.

Sukhanova, E., Tyurin, Y., Möttönen, J., and Oja, H. (2009). Multivariate tests of independence based on matrix rank and sign correlations. Manuscript.

Tamura, R. (1966). Multivariate nonparametric several-sample tests. *Annals of Mathematical Statistics*, **37**, 611–618.

Taskinen, S., Croux, C., Kankainen, A., Ollila, E., and Oja, H. (2006). Influence functions and efficiencies of the canonical correlation and vector estimates based on scatter and shape matrices. *Journal of Multivariate Analysis*, **97**, 359–384.

Taskinen, S., Kankainen, A., and Oja, H. (2002). Tests of independence based on sign and rank covariances. In: *Developments in Robust Statistics* (ed. R. Dutter, P. Filzmoser, U. Gather and P.J. Rousseeuw), Springer, Heidelberg, pp. 387–403.

Taskinen, S., Kankainen, A., and Oja, H. (2003). Sign test of independence between two random vectors. *Statistics & Probability Letters*, **62**, 9–21.

Taskinen, S., Kankainen, A., and Oja, H. (2003). Rank scores tests of multivariate independence. In: *Theory and Applications of Recent Robust Methods* (ed. M. Hubert, G. Pison, A. Stryuf and S. Van Aelst), Birkhauser, Basel, pp. 153–164.

Taskinen, S., Oja, H., and Randles, R.H. (2005). Multivariate nonparametric tests of independence. *Journal of the American Statistical Association*, **100**, 916–925.

Tatsuoka, K.S. and Tyler, D. (2000). On the uniqueness of S-functionals and M-functionals under nonelliptic distribtuions. *Annals of Statistics*, **28**, 1219–1243.

Tukey, J.W. (1975). Mathematics and the picturing of data. In *Procedings of Inernational Congress of Mathematics*, vol. 2, Vancouver, 1974, pp. 523–531.

Tyler, D.E. (1982). Radial estimates and the test for the sphericity, *Biometrika*, **69**, 429–436.

Tyler, D.E. (1983). Robustness and efficiency properties of scatter matrices, *Biometrika*, **70**, 411–420.

Tyler, D.E. (1987). A distribution-free M-estimator of multivariate scatter. *Annals of Statistics*, **15**, 234–251.

Tyler, D.E. (2002). High breakdown point multivariate M-estimation. *Estadistica*, **54**, 213–247.

Tyler, D., Critchley, F., Dumbgen, L., and Oja, H. (2009). Invariant co-ordinate selection. *Journal of the Royal Statistical Society, B*, **71**, 549–592.

Vardi, Y. and Zhang, C-H. (2001). A modified Weiszfeld algorithm for the Fermat-Weber location problem. *Math.Program.*, **90**, 559–556.

Venables, W.N. Smith, D.M., and the R Development Core Team (2009). *An Introduction to R. Notes on R: A Programming Environment for Data Analysis and Graphics.* Version 2.10.0 (2009-10-26), available at *http://www.r-project.org/*

Visuri, S., Koivunen, V., and Oja, H. (2000). Sign and rank covariance matrices. *Journal of Statistical Planning and Inference*, **91**, 557–575.

Visuri, S., Ollila, E., Koivunen, V., Möttönen, J. and Oja, H. (2003). Affine equivariant multivariate rank methods. *Journal of Statistical Planning and Inference*, **114**, 161–185.

Weber, A. (1909). *Über den Standort der Industrien*, Tübingen.

Wilks, S.S. (1935). On the independence of *k* sets of normally distributed statistical variates, *Econometrica*, **3**, 309–326.

Zhou, W. (2009). Asymptotics of the multivariate Wilcoxon regression estimates. Under revision.

Zuo, Y. and Serfling, R. (2000). General notions of statistical depth function, *Annals of Statistics*, **28**, 461–482.

Index

Lecture Notes in Statistics

For information about Volumes 1 to 144, go to http://www.springer.com/series/694

145: James H. Matis and Thomas R. Kiffe, Stochastic Population Models. viii, 220 pp., 2000.

146: Wim Schoutens, Stochastic Processes and Orthogonal Polynomials. xiv, 163 pp., 2000.

147: Jürgen Franke, Wolfgang Härdle, and Gerhard Stahl, Measuring Risk in Complex Stochastic Systems. xvi, 272 pp., 2000.

148: S.E. Ahmed and Nancy Reid, Empirical Bayes and Likelihood Inference. x, 200 pp., 2000.

149: D. Bosq, Linear Processes in Function Spaces: Theory and Applications. xv, 296 pp., 2000.

150: Tadeusz Caliński and Sanpei Kageyama, Block Designs: A Randomization Approach, Volume I: Analysis. ix, 313 pp., 2000.

151: Håkan Andersson and Tom Britton, Stochastic Epidemic Models and Their Statistical Analysis. ix, 152 pp., 2000.

152: David Ríos Insua and Fabrizio Ruggeri, Robust Bayesian Analysis. xiii, 435 pp., 2000.

153: Parimal Mukhopadhyay, Topics in Survey Sampling. x, 303 pp., 2000.

154: Regina Kaiser and Agustín Maravall, Measuring Business Cycles in Economic Time Series. vi, 190 pp., 2000.

155: Leon Willenborg and Ton de Waal, Elements of Statistical Disclosure Control. xvii, 289 pp., 2000.

156: Gordon Willmot and X. Sheldon Lin, Lundberg Approximations for Compound Distributions with Insurance Applications. xi, 272 pp., 2000.

157: Anne Boomsma, Marijtje A.J. van Duijn, and Tom A.B. Snijders (Editors), Essays on Item Response Theory. xv, 448 pp., 2000.

158: Dominique Ladiray and Benoît Quenneville, Seasonal Adjustment with the X-11 Method. xxii, 220 pp., 2001.

159: Marc Moore (Editor), Spatial Statistics: Methodological Aspects and Some Applications. xvi, 282 pp., 2001.

160: Tomasz Rychlik, Projecting Statistical Functionals. viii, 184 pp., 2001.

161: Maarten Jansen, Noise Reduction by Wavelet Thresholding. xxii, 224 pp., 2001.

162: Constantine Gatsonis, Bradley Carlin, Alicia Carriquiry, Andrew Gelman, Robert E. Kass Isabella Verdinelli, and Mike West (Editors), Case Studies in Bayesian Statistics, Volume V. xiv, 448 pp., 2001.

163: Erkki P. Liski, Nripes K. Mandal, Kirti R. Shah, and Bikas K. Sinha, Topics in Optimal Design. xii, 164 pp., 2002.

164: Peter Goos, The Optimal Design of Blocked and Split-Plot Experiments. xiv, 244 pp., 2002.

165: Karl Mosler, Multivariate Dispersion, Central Regions and Depth: The Lift Zonoid Approach. xii, 280 pp., 2002.

166: Hira L. Koul, Weighted Empirical Processes in Dynamic Nonlinear Models, Second Edition. xiii, 425 pp., 2002.

167: Constantine Gatsonis, Alicia Carriquiry, Andrew Gelman, David Higdon, Robert E. Kass, Donna Pauler, and Isabella Verdinelli (Editors), Case Studies in Bayesian Statistics, Volume VI. xiv, 376 pp., 2002.

168: Susanne Rässler, Statistical Matching: A Frequentist Theory, Practical Applications and Alternative Bayesian Approaches. xviii, 238 pp., 2002.

169: Yu. I. Ingster and Irina A. Suslina, Nonparametric Goodness-of-Fit Testing Under Gaussian Models. xiv, 453 pp., 2003.

170: Tadeusz Caliński and Sanpei Kageyama, Block Designs: A Randomization Approach, Volume II: Design. xii, 351 pp., 2003.

171: D.D. Denison, M.H. Hansen, C.C. Holmes, B. Mallick, B. Yu (Editors), Nonlinear Estimation and Classification. x, 474 pp., 2002.

172: Sneh Gulati, William J. Padgett, Parametric and Nonparametric Inference from Record-Breaking Data. ix, 112 pp., 2002.

173: Jesper Møller (Editor), Spatial Statistics and Computational Methods. xi, 214 pp., 2002.

174: Yasuko Chikuse, Statistics on Special Manifolds. xi, 418 pp., 2002.

175: Jürgen Gross, Linear Regression. xiv, 394 pp., 2003.

176: Zehua Chen, Zhidong Bai, Bimal K. Sinha, Ranked Set Sampling: Theory and Applications. xii, 224 pp., 2003.

177: Caitlin Buck and Andrew Millard (Editors), Tools for Constructing Chronologies: Crossing Disciplinary Boundaries, xvi, 263 pp., 2004.

178: Gauri Sankar Datta and Rahul Mukerjee, Probability Matching Priors: Higher Order Asymptotics, x, 144 pp., 2004.

179: D.Y. Lin and P.J. Heagerty (Editors), Proceedings of the Second Seattle Symposium in Biostatistics: Analysis of Correlated Data, vii, 336 pp., 2004.

180: Yanhong Wu, Inference for Change-Point and Post-Change Means After a CUSUM Test, xiv, 176 pp., 2004.

181: Daniel Straumann, Estimation in Conditionally Heteroscedastic Time Series Models, x, 250 pp., 2004.

182: Lixing Zhu, Nonparametric Monte Carlo Tests and Their Applications, xi, 192 pp., 2005.

183: Michel Bilodeau, Fernand Meyer, and Michel Schmitt (Editors), Space, Structure and Randomness, xiv, 416 pp., 2005.

184: Viatcheslav B. Melas, Functional Approach to Optimal Experimental Design, vii., 352 pp., 2005.

185: Adrian Baddeley, Pablo Gregori, Jorge Mateu, Radu Stoica, and Dietrich Stoyan, (Editors), Case Studies in Spatial Point Process Modeling, xiii., 324 pp., 2005.

186: Estela Bee Dagum and Pierre A. Cholette, Benchmarking, Temporal Distribution, and Reconciliation Methods for Time Series, xiv., 410 pp., 2006.

187: Patrice Bertail, Paul Doukhan and Philippe Soulier, (Editors), Dependence in Probability and Statistics, viii., 504 pp., 2006.

188: Constance van Eeden, Restricted Parameter Space Estimation Problems, vi, 176 pp., 2006.

189: Bill Thompson, The Nature of Statistical Evidence, vi, 152 pp., 2007.

190: Jérôme Dedecker, Paul Doukhan, Gabriel Lang, José R. León, Sana Louhichi Clémentine Prieur, Weak Dependence: With Examples and Applications, xvi, 336 pp., 2007.

191: Vlad Stefan Barbu and Nikolaos Liminos, Semi-Markov Chains and Hidden Semi-Markov Models toward Applications, xii, 228 pp., 2007.

192: David B. Dunson, Random Effects and Latent Variable Model Selection, 2008.

193: Alexander Meister. Deconvolution Problems in Nonparametric Statistics, 2008.

194: Dario Basso, Fortunato Pesarin, Luigi Salmaso, Aldo Solari, Permutation Tests for Stochastic Ordering and ANOVA: Theory and Applications with R, 2009.

195: Alan Genz and Frank Bretz, Computation of Multivariate Normal and *t*Probabilities, xiii, 126 pp., 2009.

196: Hrishikesh D. Vinod, Advances in Social Science Research Using R, xx, 207 pp., 2010.

197: M. González, I.M. del Puerto, T. Martinez, M. Molina, M. Mota, A. Ramos (Eds.), Workshop on Branching Processes and Their Applications, xix, 296 pp., 2010.

198: P. Jaworski, F. Durante, W. Härdle, T. Rychlik (Eds.), Copula Theory and Its Applications - Proceedings of the Workshop Held in Warsaw 25-26 September 2009, xiv, 327 pp., 2010.

199: Hannu Oja, Multivariate Nonparametric Methods with R, xii, 241 pp, 2010.